Handbook of Seal Integrity in the Food Industry

Handbook of Seal Integrity in the Food Industry

Michael Dudbridge

WILEY Blackwell

Registered Office
John Wiley & Sons, Ltd, The Atrium, Southern Gate, Chichester, West Sussex, PO19 8SQ, UK

Editorial Offices
9600 Garsington Road, Oxford, OX4 2DQ, UK
The Atrium, Southern Gate, Chichester, West Sussex, PO19 8SQ, UK
111 River Street, Hoboken, NJ 07030-5774, USA

For details of our global editorial offices, for customer services and for information about how to apply for permission to reuse the copyright material in this book please see our website at www.wiley.com/wiley-blackwell.

Library of Congress Cataloging-in-Publication Data

Names: Dudbridge, Mike, author.
Title: Handbook of seal integrity in the food industry / Michael Dudbridge.
Description: Hoboken : John Wiley & Sons, Inc., 2016. | Includes bibliographical references and index.
Identifiers: LCCN 2015048489 (print) | LCCN 2015050866 (ebook) | ISBN 9781118904565 (paperback) | ISBN 9781118904602 (pdf) | ISBN 9781118904596 (epub)
Subjects: LCSH: Food–Packaging. | Food containers. | Sealing (Technology) | BISAC: TECHNOLOGY & ENGINEERING / Food Science.
Classification: LCC TP374 .D83 2016 (print) | LCC TP374 (ebook) | DDC 664/.09–dc23
LC record available at http://lccn.loc.gov/2015048489

A catalogue record for this book is available from the British Library.

Wiley also publishes its books in a variety of electronic formats. Some content that appears in print may not be available in electronic books.

Cover image: ©mark wragg/GettyImages

Set in 9/12pt Meridien by SPi Global, Pondicherry, India

Printed in Singapore by C.O.S. Printers Pte Ltd

1 2016

Contents

Introduction

The aim of this book is to provide a useful reference for people working in the food industry who need some ideas to help solve issues they may be having with weak and leaking seals on their food packages.

The vast majority of pre-packed food is sold in flexible and semi-flexible packages that are sealed using heat. This handbook considers these packages along with other food packaging systems using rigid materials such as metal and glass. It also looks at other methods of sealing a food package such as ultrasonic, induction and cold seals using adhesive.

Seal integrity, or rather the lack of it, is a major contributor to food waste in manufacturing, distribution, retail and domestic contexts. The benefits of a fuller understanding of the causes of leaking and weak packs and the remedial actions that are required are obvious in terms of reduced cost and increased efficiency of the food supply chain, but there are also benefits in terms of food safety and brand protection.

This handbook includes a comprehensive index to allow easy access to the information you are looking for, but you can also increase your overall knowledge of the topic areas by reading whole sections and, it is hoped, answering the questions raised by your particular systems and circumstances.

I would like to thank all of the people who helped me put this handbook together, especially Proseal, Multivac and RDM, who helped with photographs and illustrations.

About the author

Mike Dudbridge has spent over 30 years managing food production operations for major food companies in the United Kingdom. When he joined the National Centre for Food Manufacturing he decided that he would focus his research work on the issue that frustrated him most during his career – the fact that problems on the sealing systems in his factories could quickly convert a high-quality, high-value product into something that could not be sold. At best the faulty pack could be opened and the product reworked, with the consequent hit on manufacturing efficiency. At worst the product, the package and a lot of time and effort would be consigned to the waste skip. By focusing on seal integrity and the performance of packaging systems Mike was able to reduce costs and improve performance in all of the factories he managed. The systems suggested in this book produce sustainable improvements in levels of waste and reductions in consumer complaints, as well as helping to improve the skills, knowledge and understanding of engineers, packing machine operators and factory management. Mike has carried out extensive work on faulty seals and their root causes, seal detection systems and methods of producing better seals. All of this work is contained in this handbook.

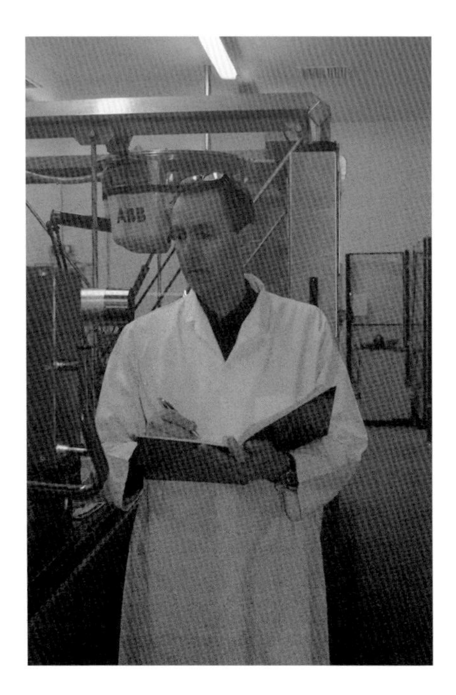

CHAPTER 1

Overview of sealing technologies, formats and systems

1.1 Introduction

This chapter provides a basic grounding in the principles of sealing a product inside a package. It includes the principles of making a seal using heat, ultrasonic energy and other sealing systems such as cold sealing using adhesives. This chapter also includes considerations of the chemistry of the materials being sealed together and how the sealing system must be designed and operated in a way that is compatible with the physical and chemical properties of the materials and also the product to be packaged.

This chapter also includes an overview of the various packaging system options available in the food industry such as bag making, pouch sealing, tray sealing, flow wrapping and form fill seal systems. These packaging options are included in much greater detail elsewhere in the book.

Finally, the chapter reviews the industry sectors that routinely use sealing technologies in their packaging processes – pharmaceuticals, food and other consumer products – and their need for pack security, shelf life extension and product protection from contamination and loss. Though the focus of the book is the food industry, the principles of sealing products into packages can be applied across many different industries.

1.2 The importance of packaging and seal integrity

Modern methods of retailing, especially food retailing, rely in the main on the shopper selecting pre-packaged products from the shelves of the retailer. It is vital that the food inside the package is protected from spoilage and leakage by the packaging that is surrounding it. Leaking packs on a supermarket shelf are an

Handbook of Seal Integrity in the Food Industry, First Edition. Michael Dudbridge.

Fig. 1.1 Modern methods of food retailing rely on pre-packaged goods being available for the shopper to select. The packaging and seal integrity are important factors in the selection process as are the distribution and safety of the foods in the shops. The extended shelf life that is offered by the packaging system is vital to the way that our food distribution systems work, making a wide variety of foods available to consumers.

indication of a lack of control in the production factory and this will cause shopper rejection of the pack and waste throughout the supply chain. If a leaking pack is inadvertently placed in the shopping basket then it may well generate a consumer complaint or in the worst case it may cause illness in the consumer.

Without robust packaging capable of withstanding the rigours of logistics and retail display the whole way that we shop for food and other consumer goods would not be possible. It is vital that factories supplying packaged goods understand the needs of the product and the needs of the consumer in terms of the package (Fig. 1.1).

There are many packaging materials, and systems and methods have been designed to use these materials to create packages to enclose, protect and retain the contents of the pack. Recent trends in packaging materials have moved away from rigid materials such as metal cans and glass jars towards semi-rigid and flexible materials such as thermoplastics. Sealing of rigid packaging is usually obtained using mechanical seals such as the double seam used in the canning industry or the screw cap in glass jars. The sealing of semi-flexible and flexible materials

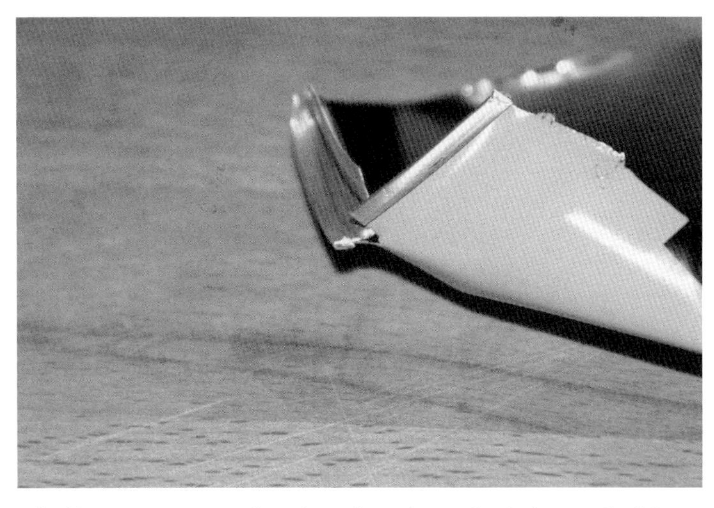

Fig. 1.2 A double seam on a canned product where the mechanical strength of the material is used to hold the parts together. The small gaps are filled with a flexible mastic compound.

requires a bond between the layers to join them together. This bond is most commonly created by the controlled application of heat to the materials, but other options are available as will be explored later.

1.3 The sealing of rigid containers

Both metal cans and glass jars rely on two factors to create a strong seal. First, there is a need of a mechanical fix to join the can to the can end or the glass jar to its lid. This mechanical fix is created by bending and forming the rigid materials after filling to prevent the can end or jar lid from separating from the body of the package. This is not enough to create a seal though. The small gaps that appear in these mechanical fixes would allow the passage of oxygen and bacteria both into and out of the pack as well as possible leakage of product. The small gaps need to be filled to ensure that contamination or leaking cannot occur. This is achieved using flexible mastics, rubber-like materials that can move and distort their shape during the making of the mechanical fix so that the small gaps can be filled. This creates a hermetic seal – one where nothing can pass into or out of the pack (Fig.1.2 and Fig. 1.3).

1.4 Hermetic seals

Hermetic seals are required when packaging foods with a long shelf life. These foods are often termed 'shelf stable' and have their long shelf life even when stored in ambient, non-refrigerated, conditions. The food inside the hermetically

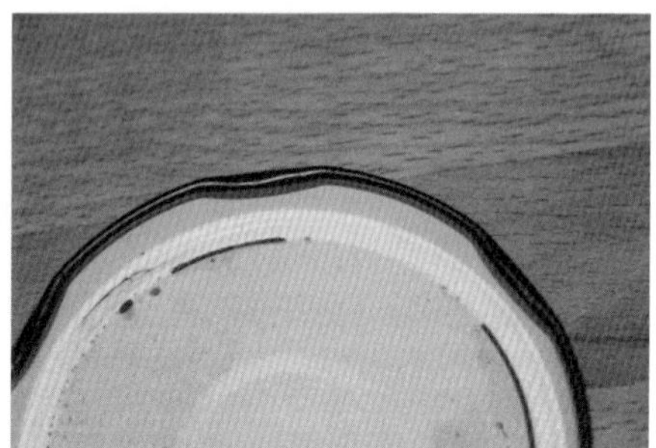

Fig. 1.3 A glass jar with a twist lid is a common package. Again, the strength is given by the materials, with the final seal being obtained using a flexible compound inside the lid.

sealed packs is often heat-processed after the packs are sealed to kill any bacteria inside the pack. This heat process is carried out in large pressure cookers, called retorts, where the packs are taken up to 120°C. This process is very demanding on the packaging. The combination of heat and pressure inside the retort means that the seals have to be very strong to withstand the process. The rigours of this process continue after the heating stage. During cooling any leak in the seal would see the can or jar suck the cooling water into the container, and with the cooling water could come bacteria and other contaminants. The development of this 'canning process' has been rigorous and the systems and package designs have been optimised to make the process as efficient as possible without risking the food safety of the products.

1.5 Developments in the canning process

In order to reduce the cost of the canning process there have been many changes. Each change is tested to ensure that the safety of the foods is not compromised as the efficiency of the process is improved. Many of the optimisation techniques used in rigid container packing systems have been repeated with semi-flexible

Fig. 1.4 Thinner materials used to reduce the cost of a package. In this can the wall thickness was reduced but the side wall strength was retained using the beaded shape, resulting in a lower cost can that performs well.

and flexible packaging system, so the themes will be seen again later in this book but applied to a different set of circumstances.

1.5.1 Thinner materials

Glass and metal are expensive materials from which to construct a food container, and as a result of this cost packages have been redesigned with less material and thinner walls, but importantly the designs retain, or even improve, the pack strength. Clever design and careful analysis has removed excess material from where it is not needed and as a result the cost of the package has been reduced without any deterioration of the pack performance (Fig. 1.4).

1.5.2 New materials

Changes in sealing mastic materials have improved performance as has the lining material on the inside of metal rigid containers. Changes here have seen improved performance especially in resisting the effects of acid foods that can attack the materials during the shelf life of the product long after the food containers have left the factory.

Fig. 1.5 The use of shrink labels adds a layer of protection to glass bottles with thinner walls.

1.5.3 Additional protection of the container from secondary packaging

Glass containers have become thinner as designers seek to reduce the package weight and this has, in some circumstances, caused an increase in damage to packages during logistics and retail display. Systems such as shrink sleeve labelling have allowed the new low-weight container to be retained because of the extra protection offered by the secondary package (Fig. 1.5).

1.6 Heat sealed packages

Semi-rigid and flexible packaging materials have the advantage over rigid packaging materials because of their low weight and low cost but they have a distinct disadvantage when it comes to sealing them to make a container. Because of their flexibility a mechanical fix for the package is not possible. To seal semi-rigid and flexible packages a different method of sealing must be adopted. The materials of the package must be bonded together to create a seal (Fig. 1.6).

Semi-rigid trays for ready meals and flexible bags for snack foods have to be sealed in order to retain the product and protect it from contamination or leakage.

Fig. 1.6 Flexible materials have been made less expensive by a process of 'down gauging' or making thinner. If this process has not been done correctly then the performance of the pack can be affected in terms of seal integrity or strength. This thinner film has had eye marks added to allow the packing machine to compensate for the extra stretch that now occurs.

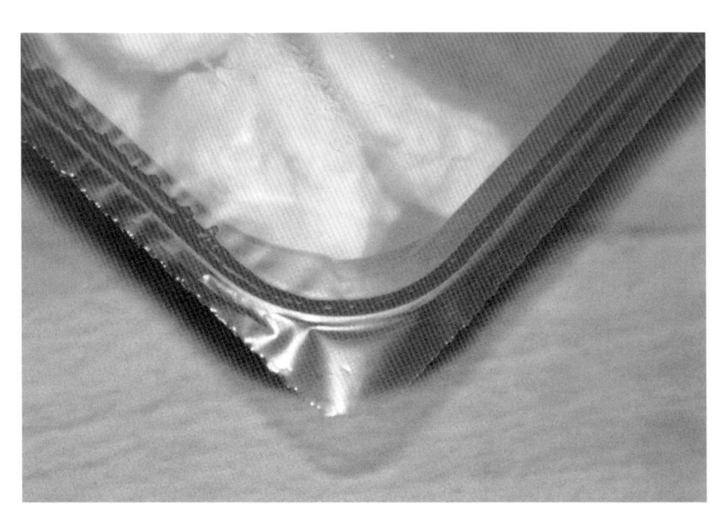

Fig. 1.7 The seals on this tray have to perform their function throughout the supply chain so the design and performance of the seals must be inspected and tested to ensure everything is as it should be.

The seal needs to be able to carry out its function during manufacture, logistics, and retail display and in the domestic situation (Fig. 1.7 and Fig. 1.8). If a seal fails at any time in the supply of the pack the product inside is open to contamination or leakage and could give rise to waste and possible consumer complaint.

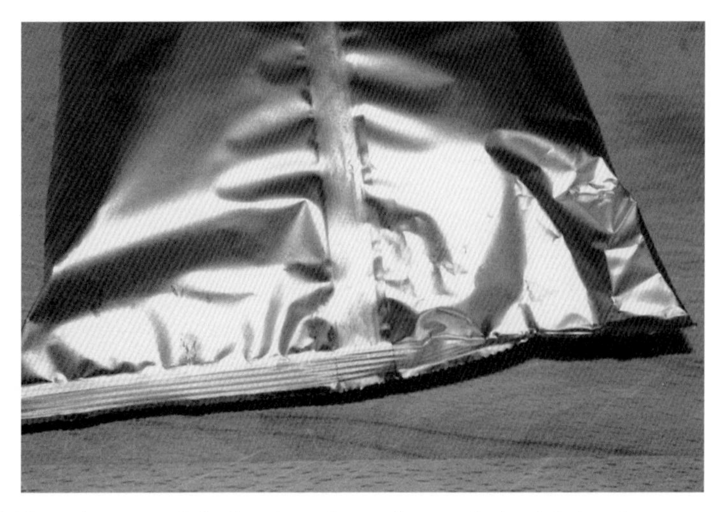

Fig. 1.8 The seals on a snack food pack need to perform to the level designed to ensure that the product reaches the consumer at the expected quality. If the seals on this pack leak then off flavours will develop and the product will go soft before the 'best before' date on the pack.

1.7 What is a good seal?

People in the food supply chain often think about creating the perfect seal. The perfect seal is one that carries out its function for the life of the product and both retains and protects the product to the desired level. Different food products have different requirements of their packaging system and so seals that may be perfect for one group of foods may be inadequate for another.

Some examples would be as follows:

In long-life, low-acid ambient food products such as a can of soup the need of the seals on the can is that they are hermetic and do not allow the passage of air or bacteria into the product. If a can seam is examined under a microscope it may well be possible to find very, very small leaks of the order of 2 or 3 microns (millionths of a meter). The leaks are smaller than a bacterial cell so none could pass through and the rate of diffusion of air into the container would be so slow that it would be insignificant in terms of the shelf life of the product (Fig. 1.9). Indeed, the rate of oxygen migration into the pack may well be faster through the packing material itself. The so-called OTR (oxygen transmission rate) for packaging materials is a measure of their resistance to the flow of oxygen molecules through the packaging material itself.

In a pack of biscuits the primary function of the packaging is to collate and retain the product and reduce the rate of moisture transmission into the pack from the environment. The shelf life of the product may be similar to that of a can of soup and this too will be stored and displayed at ambient temperatures. Biscuits are low in moisture and high in sugar and would therefore be classed

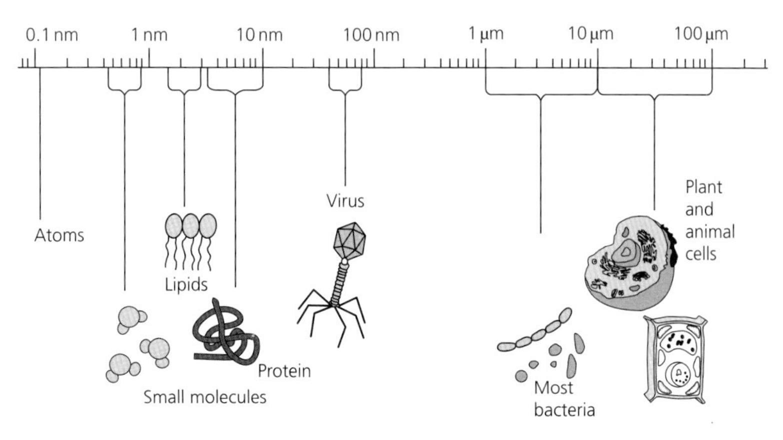

Fig. 1.9 Very small 'micro' leaks in the seal of a can or food package may be smaller than a bacterial cell and cause no risk to the product or its consumer from bacterial contamination, but as shown here, there are other considerations in the packaging of food products when you consider microscopic leaks.

as a low-risk food with respect to the potential for food poisoning. Because of the type of food and its requirements, it is clear that the package and its seals can be very different from those for a can of soup. Biscuits need protection from the ingress of moisture into the pack. On every pack of biscuits will be the storage instructions 'store in a cool dry place'; this is to protect the product from the effects of moisture in the air. Leaking seals will not help the storage of dried products, but the packaging material itself will allow moisture through. The MVTR (moisture vapour transmission rate) for packaging materials is a measure of their resistance to the flow of moisture through the packaging material itself (Fig. 1.10).

In a pack of ready-to-eat cured meat the package may have been created with a modified and protective atmosphere inside to extend the shelf life of the product. Seals here not only have to retain the product and prevent bacterial contamination, they also have to retain the modified atmosphere inside the pack (Fig. 1.11). The important gasses in these systems are carbon dioxide and oxygen. The carbon dioxide needs to stay inside the pack to retain its protective effect and the oxygen needs to be kept out to prevent it facilitating chemical changes and encouraging the growth of bacteria already on the product when it was packed. It can be seen that the requirement for good seals in this scenario is paramount if the food is going to be preserved and be safe to eat. Preventing the migration of carbon dioxide out of the pack is very difficult because of the small molecule size, and preventing the migration of oxygen into the pack is even more demanding as the molecules are even smaller. The partial pressures of these gasses will encourage migration with even the smallest of faults in the seal. So in high-risk modified atmosphere packs the need for a very high level of seal integrity is vital.

Fig. 1.10 The moisture vapour transmission rate will have implications for the quality of dry foods. The packaging material as well as the seals will allow the passage of moisture.

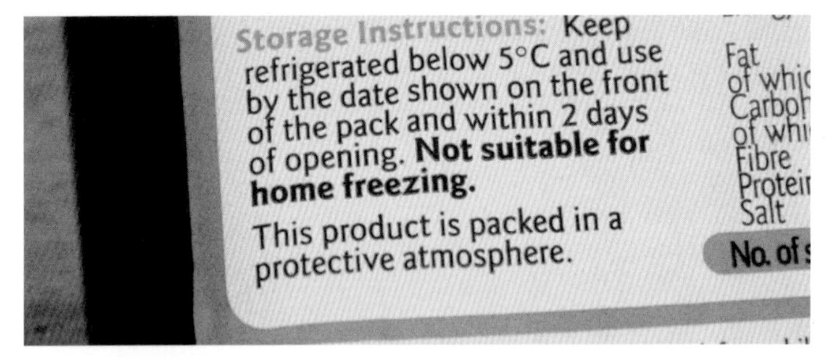

Fig. 1.11 The use of modified atmosphere packaging (MAP) and vacuum packaging puts very high demands on the integrity of the seals in the pack. Poor seal integrity would lead to gas exchange with the atmosphere outside of the pack and possible consequences for the shelf life and safety of the product.

1.8 So ... what is a good seal?

It can be seen from the examples above that a good seal is one that is adequate for the job it is required to do. For factories it therefore becomes important that decisions are made on the specification and performance of the seals it is making in its packages. A testing procedure for the seals should be set up to ensure that the factory is not producing seals that are inadequate in terms of their integrity or strength. Finally, the factory needs to set up management and control systems to ensure that if an inadequate seal is made it can be detected and that it is not allowed out into the supply chain (Fig. 1.12).

Company XYZ

Seal Inspection Sheet

)ate

/lachine number

nstructions - three samples should be taken from the machine at the start and end of the production run and also every fifteen minutes during the run.
'he seals should be inspected and the machine settings recorded.

Time	Product	Seal quality			Temperature 1	Temperature 2	Temperature 3	Dwell Time	Corrective Action	Signiture
		Seal A	Seal B	Seal C						

End of Shift Sign Off

Fig. 1.12 A typical seal inspection sheet used to collect data from seal inspection tests carried out during packaging operations. Notice the frequency of the tests in this example. Is it good enough to inspect such a small proportion of the total number of packs when the quality and safety of the product is so vital? In certain factories, where seal integrity is a vital part of the safety of the product, the hazard analysis critical control points (HACCP) system should reflect this.

1.9 Seal management in a packing operation

In order for a factory to consistently produce seals that are adequate for the requirements of the product inside the pack, adequate for the logistics operation in the supply chain and adequate for the retail and domestic use and display of the product, a seal management system needs to be operated. A correctly designed seal management system will ensure that the packages produced meet the required standard in terms of seal strength and integrity and that the procedures in the factory are robust and able to detect a problem when it arises.

1.9.1 Setting the required standards

The first step in seal management is to ensure that clear standards are set for the performance and testing of the seals being produced. Typically, this means that performance standards are set within a defined and standardised test. International standards have been developed for the testing of heat seals and these are a good starting point, especially if the business has more than one factory, but working to standards set by ASTM International (formerly the American Society for Testing and Materials) can be good practice and ensure that your seals meet an internationally recognised standard.

1.9.2 Second step – set up a testing system for seals in your business

Calibration of testing equipment is always a good initial step in developing a testing procedure. We will see in Chapter 5 that this is not always carried out in a robust way. The use of the 'manual squeeze test' for checking packs is always an inadequate test. There needs to be some actual measurement here if good seal performance is to be obtained. Once a testing system has been developed, the standards of what is acceptable can be set. Typically a seal test would consist of a measure of the seal strength – maybe a peel test where the amount of force required to peel the seal apart is measured – and a test of the seal integrity – a leak test where perhaps the pack is placed in a vacuum tank and held under water (Fig. 1.13; the operator of the test is looking for bubbles from the pack), the pack is inflated and a pressure decay looked for or the pack is cut open and a dye is placed inside the pack (if the dye leaks out through the seals it indicates a problem). All of these tests are 'off line' and are maybe carried out every 15 or 20 minutes on single packs. The tests are also destructive in nature so not every pack can be tested. There is an increasing amount of attention being paid now to 'on-line' 100% inspection systems for the non-destructive testing of seals. The 100% testing systems fall into four groups.

Fig. 1.13 Typical leak testing equipment to establish if the seals being made meet the required performance. There are many types of testing equipment available. Important is that the tests are carried out in a standard way, that the equipment is calibrated and that corrective action is taken if the results of the tests indicate that there may be a problem with the seal strength or integrity. (Pictures by kind permission of RDM Test Equipment Ltd.)

1.10 Testing systems

1.10.1 Mechanical squeeze test

This is where the pack is squeezed mechanically and a movement in the pack is detected by the testing machine. This can be done using a converging roller system above and below the pack or with a ram system where a flat plate is pushed into the pack and the resistance of the pack is measured. If the resistance of the pack to the push declines then it indicates a leaking pack.

1.10.2 Sniffer systems

These are where a tracer gas is placed inside the pack when the pack is sealed and then the presence of the gas outside the pack after sealing indicates a leaking pack. The trace gas is typically carbon dioxide, which is used in modified atmosphere packaging systems, but other gasses have also been introduced into packages with the sole purpose of using them to help detect leaks. This area will be explored further in the final chapter of this book, which looks at innovations and new developments.

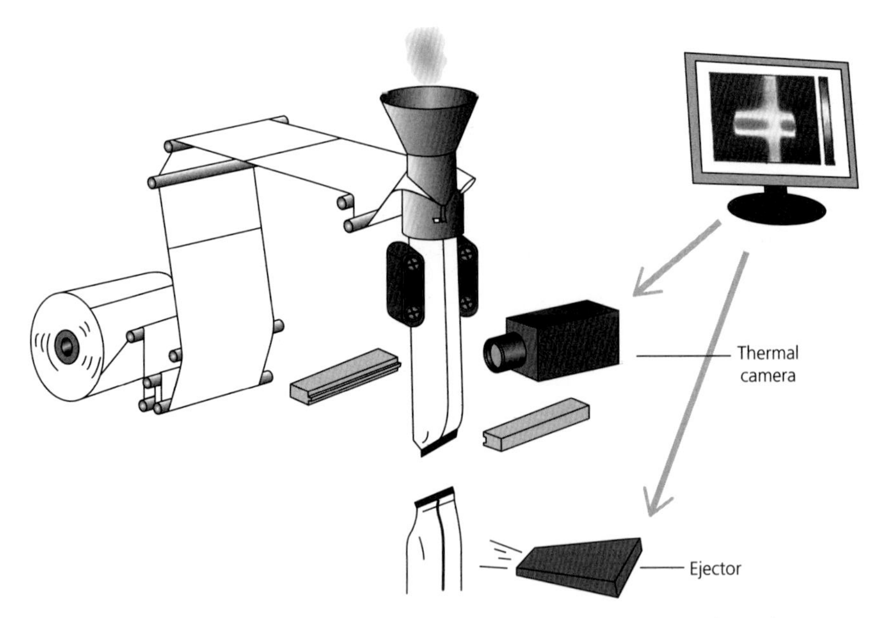

Fig. 1.14 A typical camera system automatically inspecting seals on the exit of a packaging machine. (Picture by kind permission of RDM Test Equipment Ltd.)

1.10.3 Vision systems

These are where cameras mounted above the out-feed conveyors of a sealing machine capture images of each pack and then a computer analyses each image looking for indications of anomalies in the seal area (Fig. 1.14). The vision system can use normal light to inspect the seals or it can use X-ray or the infrared spectrum to carry out the inspection.

1.10.4 Ultrasound systems

These are where the echo bouncing back from a pack is analysed allowing for seal integrity issues to be detected. The ultrasound creates an image of the seal and can therefore be used to detect anomalies in the seal area. Image capture time and image processing speed are improving, so while initially the system was used off line, it will soon be able to match packing speeds, allowing it to be applied in real time.

All of these on-line systems are subject to error and will generate false positives (saying there is a problem when the seals are fine) and false negatives (not detecting a seal fault when one is present), so the systems need to be calibrated and regularly checked to make sure they are working correctly. Where 100% inspection systems are used, the checking and calibration of them forms a very important part of the seal management in the factory.

1.11 How is a good seal made?

The sealing of flexible and semi-rigid packages can be carried out in a variety of ways. Here are the basic principles of some of the more common methods.

Fig. 1.15 Heated bar sealing. Perhaps the simplest form of generating a heat seal, a heated bar is pressed onto two layers of a thermoplastic and after a short dwell time of around half a second a seal is made. When the heated bar is removed there needs to be a short cooling time of around a quarter of a second to allow the thermoplastic materials to set in their new sealed configuration before any pressure or stress is put onto the seal. In this picture the heated part of the sealing system is covered with a layer of polytetrafluoroethylene (PTFE) tape to stop the thermoplastic from sticking to the heated surfaces of the sealing jaws.

1.12 Heat sealing

This is by far the most common method. Two surfaces of a thermoplastic material are brought together and heat is applied to the material. The thermoplastic material softens and the long chain molecules of the two surfaces join together. Depending on the type of thermoplastic, this joining can be either a chemical bond or a physical interweaving. Heat can be applied in many ways to the packaging materials.

1.12.1 Heated tooling

This is where a pre-heated bar or tool is pressed against the outside of the materials to be sealed (Fig. 1.15). Heat is transferred by conduction to the touching surfaces where the seal is made. This process is quite low cost and straightforward. It does, however, suffer from the slow transfer of heat through the materials, and this is especially important with thicker materials. The heated bar temperatures and the time that they are in contact with the outer surface of the materials can also lead to distortion of the package with some thermoplastics. The heat in the packaging materials also has to be dissipated after the seal is made to prevent the seal opening again because of the continued fluidity of the thermoplastic materials while they are still at elevated temperatures.

1.12.2 Induction sealing

This is a system often used to seal caps onto plastic bottles. A magnetic coil is used to create a magnetic field that induces an electrical current in a foil layer of the cap. The induced electrical current produces resistive heating in the foil layer and

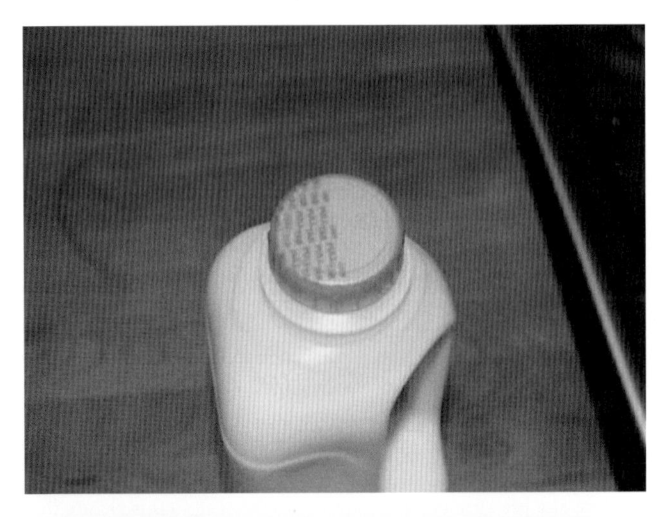

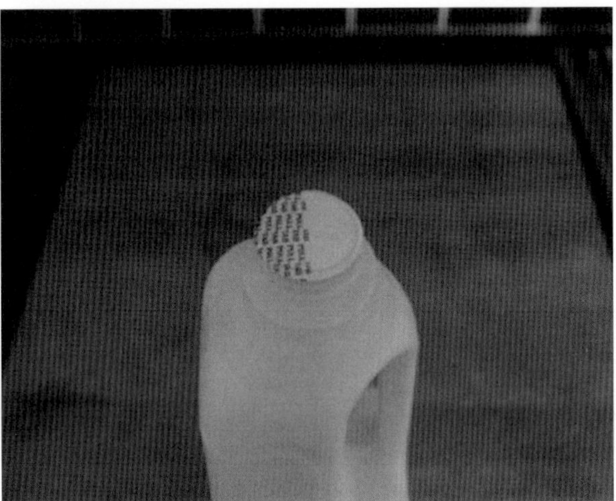

Fig. 1.16 An induction sealing system where a magnetic field induces an electrical current in a thin layer of foil. The subsequent heating is used to seal two thermoplastic layers together. This system is typically used in the sealing of plastic milk bottles.

this heat is then passed to a thermoplastic layer of the cap. The heat is then sufficient to soften the thermoplastic and cause a seal to be made (Fig. 1.16).

1.12.3 Ultrasonic sealing

Sound in the frequency between 20 and 40 kHz is used to create friction between two layers to be joined. The ultrasonic vibrations generate heat in the material surfaces so that a seal can be made. Energy is introduced into the seal area using an ultrasonic horn and anvil which causes local rises in temperature resulting in the seal. The vibrations have also been observed to move seal area contamination out of the way, so this method is claimed to be better than conventional heat sealing at forming reliable seals in circumstances where a lot of seal area contamination occurs.

1.12.4 Spin welding

Here friction between two layers to be joined is used to generate heat and therefore a seal. The two parts to be sealed are pushed together and spun in opposite directions to generate the heat required. By definition, here the parts being joined have to be circular. This system is used, for example, to seal together the two halves of the widget that is used in the packaging of some canned beer.

1.12.5 Hot gas/radiant heat sealing

Here the surfaces to be sealed are heated by either hot air or radiant heat and then, when the thermoplastic has softened, the surfaces are pushed together. This system is often used with thicker packaging materials where waiting for the heat to conduct through the material is not an option. Packing that is made up of multilayer laminates that include cardboard are often sealed this way. If the heat was applied to the outside of the material then the outside would be damaged before the inside had reached a sealing temperature. Drinks cartons and sandwich packaging can be sealed in this way.

1.13 Non-heat sealing methods

Other non-heat methods of sealing are also employed and should be included here for the sake of completeness.

1.13.1 Cold sealing

This is a method of sealing a package that relies on the use of a latex cohesive layer applied to the packaging material. The surfaces are pressed together and the cohesive latex seals them. This is typically used with confectionery and ice cream products where the application of a heat sealing process may cause quality issues with the product (Fig. 1.17).

1.13.2 Adhesive sealing

In some applications an adhesive is used on one of the surfaces to make a closure. The adhesive is typically a hot melt and is used to seal cardboard cartons in the frozen and chilled food industry where perfect seal integrity is not required.

1.13.3 Solvent sealing systems

In these systems a solvent is used to break the intermolecular bonds in the surface of the plastic materials. Once the solvent is applied, the surfaces are pressed together and the molecules form new bonds with molecules from the other

Fig. 1.17 The cold seal latex adhesive on the seal areas of a confectionery pack.

surface. This molecular bonding technique is very specialised but is used in some aspects of product packaging.

1.14 Packaging materials

So far we have briefly introduced sealing systems and some of the background of packaging. One important aspect of packaging systems is the packaging materials. There is a vast range of packaging material types, so first we will look at the important parameters when considering this aspect of sealing. We look in detail at materials later in the book, but here is an introduction to some of the factors that need to be considered.

1.15 Sealing parameters

There are several characteristics of packaging materials that need to be considered when selecting the best solution to match a need.

1.15.1 Sealing temperature

Thermoplastics of different chemical make-up and different structures have different melting points. The melting point of a thermoplastic is the point where the intermolecular bonds are reduced and the plastic takes on the behaviour of a liquid. In packaging we want the thermoplastic to start to loosen the intermolecular bonds to facilitate the sealing process but we do not want liquid behaviour.

The sealing temperature is one where the intermolecular bonds in the thermoplastic loosen and molecules at the surface of the material are able to intermesh or chemically bond with molecules from the surface of the other piece of packaging material. Sealing temperature is really a small range of temperatures over which the desired seal can be made without the thermoplastic distorting or becoming so heated that the structure of the material is fundamentally changed. We will look again at the thermal properties of packaging materials later in the book as an understanding of packaging material behaviours is important.

1.15.2 Flow characteristics

When seals are made the softened surfaces of the thermoplastics are pushed together to give a good contact and the opportunity for the required intermolecular entanglement and chemical bonding to occur. If the seal area is contaminated then the two surfaces will not be able to join together. Some packaging materials have been designed with high-flow thermoplastics at the sealing surface. The plan with this type of design is that the thermoplastic material (in its semi-liquid state) will flow as the materials are forced together. The flow of the thermoplastic is designed to move seal area contamination away from the seal area and allow a good seal to be made. Packaging materials are designed to have certain sealing and flow characteristics by the manipulation of the chemistry and structure of the material. Often different types of thermoplastic materials are blended to give the required melting and flowing properties. These co-polymers mean that packaging materials can be designed to exactly meet the requirements of the business using them. Obviously, the more specialised the requirement the more expensive is the material but it is possible to fine tune the packaging materials to meet the needs of the business. Where more flow is required it is possible to increase the thickness for the material in a multilayer laminate to give more moving thermoplastic material and as a result sweep more contamination from the seal area. Where the flow of the thermoplastic is too great (maybe due to excessive dwell time) there is a quality fault called 'angel hair' where the thermoplastic exudes from the material and adheres to the sealing jaws. As the jaws pull away from the pack at the end of the dwell time the exuded thermoplastic is drawn out into very thin strands that look like hair. If there is an excessive or unusual quantity of angel hair on the jaws of your system it indicates excessive temperatures or dwell time.

1.15.3 Surface printing

Materials exhibit different qualities for printing on. Some materials accept inks very well and are able to hold a high-definition multicolour image. Other materials produce printing results that are not so good. Inks can migrate through packaging materials and this needs to be understood and defined when selecting packaging materials. The final part of printability of packaging is where the print occurs in the same area that a seal is going to be made. The impact of the ink on the sealing process needs to be understood as often when heated tools are used to form a seal on a printed surface the ink can transfer to the tool and build up over a period of time. This can have the impact of reducing the heat transfer to the seal areas and ultimately can cause the seals to become weaker and fail.

1.15.4 Material strength

Different packaging materials have different strength characteristics and resistance to damage. Some thermoplastics are quite stretchy at room temperatures but become less so in a chilled or frozen environment. Some materials exhibit very good puncture resistance and others less so. There is a particular property of thermoplastics that needs to be considered: tear propagation. This concerns whether, when the sachet is opened by tearing across the top, the tear goes in a straight line or whether the propagation is random such that it could lead to the contents of the pack being wasted. This property can have big implications for the design of the packing system as well as for the accessibility of the pack to the consumer. The strength of a pack needs to be adequate to withstand the handling it will receive in distribution and retail display. Sometimes the pack material is strong enough before sealing but after the heating and cooling the material strength can change and as a result the material can become weaker and more liable to fracture.

1.15.5 Material structures

There are a large number of multilayer packaging materials. These are sometimes called laminates and they consist of different layers each of which brings its own characteristics to the packaging material. Some layers add strength and puncture resistance, other layers bring an oxygen barrier and, finally, some layers are put into the laminated material to impart sealing characteristics.

There has been a recent trend towards single-layer packing materials to improve recycling, and these materials have to carry out all of the required functions. These are often called mono-layer packaging materials.

With respect to sealing characteristics, there are some important parameters that need to be considered and the choice of materials used is based on the performance of different thermoplastics within these parameters (Fig. 1.18).

Material	Clarity	MTR	OTR	Impact strength	Tg	Tm
PET	Excellent	2	75	Good	80	250
HDPE	Poor	0.5	4000	Good	−30	135
PP	Poor	0.5	4000	Fair	−20	165
PS	Excellent	10	6000	Poor	100	Amorphous
PLA	Very good	22	42	Good	55	180

Fig. 1.18 Table of typical packaging thermoplastics with some indication of their performance in different parameters. The use of polymer mixtures can change these performance numbers radically so this table is intended as a guide only. More details of packaging materials are contained in Chapter 7. Tm is the melting temperature but the material does not have to be melted to create a seal. Tg is the so-called glass transition where the material starts to become more flexible as the intermolecular structures break and flex.

1.16 Packaging systems

Packages are made in many different ways and a wide variety of technology is used to improve packaging system performance. Some believe that high-speed packaging machines are the ultimate in interaction between a machine and a material, and many design solutions have been developed to produce the highest performing system.

1.16.1 Bag-making systems

Sometimes called vertical form fill seal (VFFS) systems, this is the single most popular way of pre-packaging products in the world. This is the type of system used for the packaging of snack foods, confectionery, frozen vegetables and breakfast cereals. A packaging material is delivered to the packing factory in the form of a reel of plastic film. The film is formed into a tube and the back seal of the bag is made. Next, product is delivered down the tube and sealing jaws close to seal the top of the bag. Simultaneously, the sealing jaws seal the bottom of the next bag and cut the film to release the filled bag. These machines can run very quickly at speeds close to 200 packs per minute on certain products, and so the time available to make a seal is very short (Fig. 1.19). As a result of the short time available it is important that the materials being sealed (the inside surfaces of the packaging film) have the correct characteristics to make a good pack with adequate seals. If the inner, sealing, surfaces are made of the wrong materials the packaging speed will be slowed down and reject packs will be made. One possible response in these circumstances is to increase the temperature of the sealing jaws to drive more heat into the sealing area. This may well result in seal area distortions and the pack looking scruffy and overheated.

1.16.2 Pouch sealing systems

This is a system often used in the creation of food pouches for pet foods or microwave rice packs. These products are both long life and displayed at ambient temperatures so seal integrity is very important to the quality and safety of the food. Often in these systems three sides of the pouch are made and sealed in a separate process (often in another factory or even in another country). The preformed pouch is fed into the filling machine, opened, filled and then the top seal is made. These products are then further processed using retorts to cook the contents of the pouch. Again there are many forms of machine to fill and seal pouches but they all have the same need to fill the pouch without contaminating the seal area. The pouches are usually made of quite thick materials (often multilayer laminates) and so the heating of the thermoplastic to make the seal can take some time. The seal also retains its heat after sealing so often there is a need to cool the seal area to return the thermoplastic layer to a more stable temperature before any pressure is put onto the seal.

Fig. 1.19 A typical VFFS bag-making machine. This one is a large-scale machine capable of packing 5 kg of potatoes at around 40 bags per minute. Other machines are used for high-speed packing of snack foods with speeds approaching 200 bags per minute.

1.16.3 Tray sealers

This is a system where a preformed tray is sealed with a thin film lid (Fig. 1.20). It is a very popular packaging format in the ready meals industry. There is a difference between this system and the previous one. In this system we have two materials of different thickness being joined together and this presents some issues when it comes to heating the seal area. In this system the seal area tends to be heated from only one side and this has an impact on the temperatures reached and the speed of operation. A cycle time on a machine to seal trays would be typically 4 seconds with a seal time of around 1 second. To increase the output of the machine to higher numbers of trays per minute, machines have increased in size and are sealing multiple trays at the same time. The bag maker and the pouch sealer typically seal just one package at a time.

1.16.4 Horizontal form fill seal (HFFS) systems

This is a system where the tray is thermoformed on the machine rather than the tray being preformed in a supplying factory (Fig. 1.21). These systems are widely used for the packaging of sliced meat and cheese products. The same

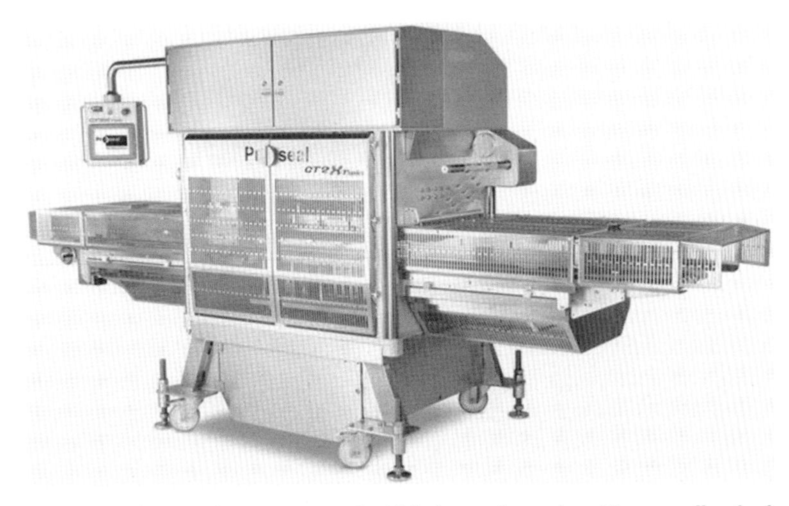

Fig. 1.20 A typical tray sealing system used widely for ready meal packing as well as fresh poultry portions. (Picture by kind permission of Proseal Ltd.)

Fig. 1.21 A typical HFFS used widely for the packaging of sliced products including ham, bacon and sliced cheese products. (Picture by kind permission of Multivac UK Ltd.)

considerations apply here in terms of sealing the packs as apply to preformed trays. The heat is typically applied from one side only so time is needed for a good seal to be made. The rate-limiting step on a thermoforming machine is typically the forming of the trays.

1.16.5 Flow wrapping system

This is a high-speed system used in many parts of the food industry to contain and protect products. The system is most popular with solid single items like confectionary bars, meat pies and even pizzas but has also been developed using trays and backing boards to be able to wrap multiple items such as crumpets, biscuits and sausages. The latest innovation in this system of packaging has been to use it to wrap minced beef in a modified atmosphere to help reduce the packaging weight of the traditional packing method of a sealed tray.

There are other packaging formats that will be covered later in the book, especially those that are common in the dairy and beverage industries.

All packaging formats and systems that rely on the effect of heat on a thermoplastic to generate a seal (no matter how that heat is applied or generated) suffer a similar set of issues during the formation of the pack that can compromise the integrity or strength of the seal. We will look in detail later at the major causes of seal integrity issues but suffice it for now to point out that on high-speed packing machines, using defined thermoplastics and with defined machine settings and tolerances, it does not take much to disturb the interaction between machine and packaging material and as a result a faulty pack can be made.

1.17 The requirements of industry for seal integrity in its packaging systems – an overview

1.17.1 Safety

The integrity of a pack is paramount to considerations of food safety. The risk of product contamination with bacteria or other substances is a huge consideration in many markets but especially for food and beverages. Poor seal integrity can result in consumer complaint or illness with a consequent negative impact on brand image. Loss of business and bad publicity are strong possibilities if seal integrity is not managed correctly within a food manufacturing business.

1.17.2 Containment

The prevention of leaks from packages is an obvious and visible requirement in distribution chains. If product is able to escape from a package then the package becomes unsellable and there is also the knock-on effect on neighbouring packages causing them to become unsellable too (Fig. 1.22).

1.17.3 Shelf-life extension

Some products can retain their quality and safety for longer if the package is fully sealed and seal integrity is good. Snack foods are often packaged in a low-oxygen atmosphere to reduce rancidity and flavour changes in the product over time. If the package is not fully sealed then the protective atmosphere inside the package may leak out leaving the way clear for the ingress of oxygen and moisture. As a result, the snack food may become stale more rapidly than expected and become unsellable.

Without the modified atmosphere being correctly sealed into the pack its protective effect is quickly lost, so the product is likely to be of lower quality or even dangerous at the end of the shelf life.

Shelf-life extension is also a feature of non-MAP (modified atmosphere packaging) systems. The correct protection in an adequately sealed package can extend

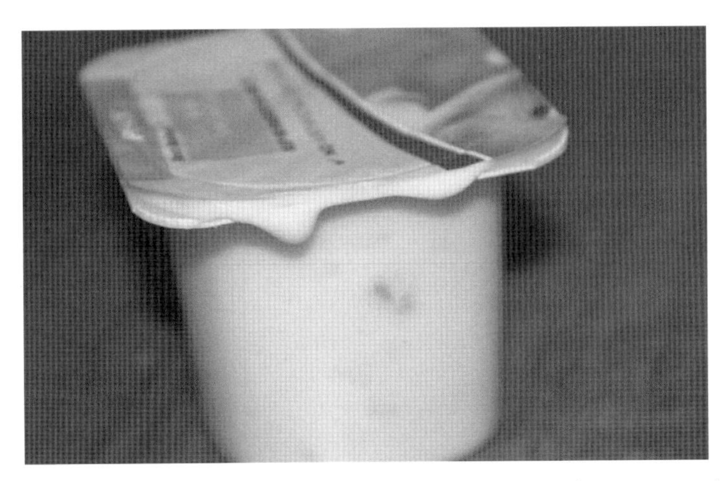

Fig. 1.22 A leaking yoghurt container will spoil the rest of the pots in the case, so one leaking container can cause more waste than just that one container. Sometimes whole pallets of product are wasted because of spillage from one container. This is especially the case with something like vegetable oil or beetroot, for example.

the life of everything from a cucumber, by reducing moisture loss and slowing respiration, to biscuits, by reducing moisture pick-up.

1.17.4 Seal strength

A seal needs to have sufficient strength to withstand the rigours of logistics and retail display. A weak seal leaving your factory could fail and as a result cause an issue in the supply chain even though the pack was sealed when it was made. There is a compromise between seal strength and the openability of a package and this is why seal strength measurements should be undertaken as well as seal integrity measurements (Fig. 1.23).

1.17.5 Accessibility

One of the major sources of consumer complaint to retailers is customers who are unable to gain access to the contents of a package without having to attack the pack with sharp implements (Fig. 1.24). Injuries caused by inaccessible packaging are a major concern, especially as the population in the developing world gets older and dexterity becomes a bigger problem. Packages are being designed with easy-open features to try to help with the balance that is required between seal strength and accessibility. Accessibility is especially an issue with meals that are reheated in the home inside the packaging. Peeling the lid from a tray of curry as it is taken from the microwave oven can be difficult enough, but a real safety issue can occur if the lid of the tray is firmly attached to the tray. The potential for burns and scalds is a risk that should be taken into account in packaging design and in the operation of the sealing systems. It is important to make sure the pack performs as required when it is in the hands of the consumer.

Fig. 1.23 A typical method of measuring seal strength using a peel test. The force required to peel the seal apart is measured to check it meets the requirements of being strong enough to pass through logistics and retail sale while at the same time allowing the consumer to open the pack. (Picture by kind permission of RDM Test Equipment Ltd.)

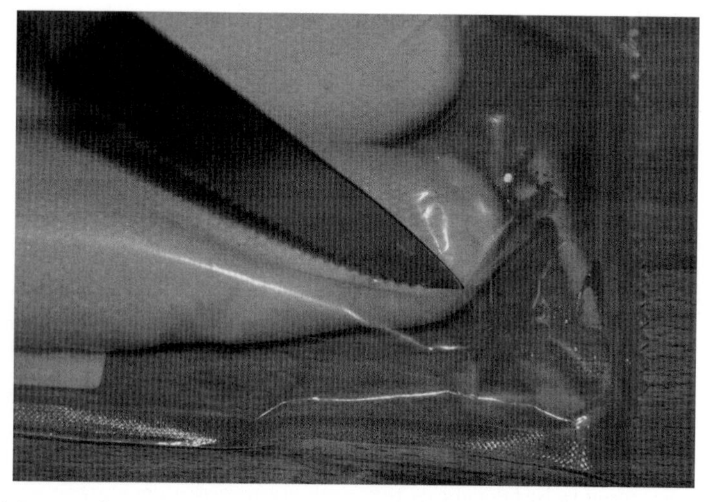

Fig. 1.24 Frustrated consumers complain in large numbers if the packaging is not correct on the products they buy. Too weak and leaks occur on the way home from the shop. Too strong and injuries can be caused trying to get into the pack.

1.17.6 Display attractiveness

With modern retail methods and the competition for shelf space the attractiveness of a package on the shelf can be the difference between making a sale or not for your product. A well-designed pack can be rendered unsellable by a poorly formed seal. The seal could be perfectly sound in terms of its strength and integrity but if it

Fig. 1.25 Here is an overheated seal on a product. It looks scruffy as the thermoplastic materials have been made too hot. As a result, as the thermoplastic cooled it formed crystals in its structure making the material hard and brittle. This is a common problem in some industries where poorly trained packing machine operators feel that the correct response to poor seals is to turn up the temperature of the heat sealing system – it's their 'go to' response to weak seals. The true cause of the sealing problem is never investigated and as a consequence the root problem is never eliminated. This picture also shows that a dye that has been placed inside the pack is starting to leak out in the centre of the pack. So the high temperature has not solved the leaking pack issue.

looks like there has been a problem with the package there is a chance that the pack will not be selected by the shopper (Fig. 1.25).

1.17.7 Portion control and multi-compartment trays

Some pack designs require good seals in order to control portions and doses. Multi-chamber packaging with individual portions is a growing trend to try to assist the consumer in the use of the product. The seals made on a pack are not always just around the perimeter. They are sometimes required to keep the components of a product separated until the point of opening and use. For example, yoghurt may need to be kept separated from crunchy pieces until the consumer is ready to eat the product; a sauce may need to be kept separated from rice or pasta in a ready meal. Checking the seal integrity of the internal seal will be impossible without destructive testing, so techniques need to be developed to allow routine checks to occur and to ensure that all is as it should be.

1.17.8 Consumer confidence

Pre-packaged products are purchased by consumers partly because they have confidence in what is inside the package. They know that their product has not been tampered with and it has been sealed to give them that confidence. If a package is discovered to be not sealed correctly then all confidence in the contents of the package can be lost. If pack seals are not adequate the product is left liable to damage and deterioration, and the consumer is aware of this. Sealed packages are a major feature of modern retail systems for a wide variety of products from food to DIY and from pharmaceuticals to consumer electronics. The seal is a major part of what is being purchased in the eyes of the consumer. Without the seal being in place the consumer will treat the product inside the package with some level of suspicion. A consumer is buying more than just the contents of a package; they are buying confidence in the contents, and good seals are and important aspect of that customer assurance.

1.18 Industry sectors – some recent increases in seal performance requirements

Pharmaceuticals and foods are the major consumer goods sectors where seals are important to the physical, chemical and biological safety and the quality of the product in the pack, but packaging seals are also important in the very function of some products. Innovation in the coffee market has brought forward a group of products that could be described as coffee pods. Without seals of exacting specifications the pods would not function correctly and the product would not be possible at all. The creation of multi-chamber washing detergent pods has a similar requirement. The phased release of the washing chemicals during the wash cycle could not work without correct sealing both internally and externally. The creation of new packaging formats and new packaging materials often hinges on the basic requirement of creating a reliable and predictable seal. There have been recent examples where new packaging materials were developed which were fully compostable so that they would break down in landfill or anaerobic digestion systems and would be more environmentally friendly as a result. The new materials proved difficult to use because the packing material was not as tolerant of temperature variations on the sealing machines. Slightly too cold and a seal was not made. Slightly too hot and the same thing happened. To use this type of material new control systems had to be developed for the sealing machines to keep the sealing temperatures in a much tighter tolerance than had previously been required with conventional thermoplastics. Some products rely on good seal integrity for their very existence. Without good reliable seals there are a number of product groups that could not exist. Equally, there are product groups for which seal integrity is so vital to the shelf life of the product that the economic viability of that whole category would not work. Waste would greatly increase and on-shelf availability would decrease; consumers would be unwilling to pay the higher prices that would result. An example here is the supply of sliced cured ready-to-eat meat products. Without good seal performance the shelf life of the product is considerably reduced to the point of being unviable.

1.19 Making changes to packaging materials or systems

It is vital that if any changes are planned with either a packaging material or a packing machine that the implications of the change are fully explored. It is possible that even a minor change could have unpredicted consequences and as a result put consumers and company reputations at risk.

For example, a simple move to make some packaging thinner to reduce the waste implications when the product is used – called 'light-weighting' – can have implications for the packaging system where its performance could change because of the thinner gauge materials being used. It could also have implications for the shelf life of the product, with increased oxygen and moisture transmission rates. As a result of the unpredicted changes, the costs of the increasing waste from the packing operation, the supply chain and the consumer could exceed the saving made in packaging weight. So any change in packaging should be fully assessed with a complete analysis of all the possible implications before the change is confirmed. Making changes to packaging systems has to be carefully considered to ensure that the seals, a vital component of the package, are not impacted by any changes. This lesson was learned during changes for the fresh produce industry. Many companies discovered difficulties in sealing PLA (polylactic acid) packaging materials when changes were made. The reason for the change was that PLA was seen as being compostable material and beneficial to the environment, but the change caused an initial large increase in sealing problems and consequent increase in waste.

Even a change of supplier has to be carefully tested to ensure a predicable performance is achieved. Thermoplastic materials from different suppliers will have different packaging performance even for materials that are called the same thing. There are wide variations in material performance and temperature/time requirements for something that is notionally the same material. This is caused by slight differences in processing and storage conditions at different suppliers. So while the chemistry says it's the same material, its ability to produce the same seal characteristics is not the same. This is especially the case where recycled materials are used in a package. As the proportion of recycled materials is increasing there is a tendency for the variability of sealing performance to increase also.

CHAPTER 2

The operation and design of tray sealing machines

2.1 Introduction

Sealing of products into preformed trays is a popular method of protecting, preserving and presenting products, and the technology involved in the creation of these packs is quite varied. The machines designed to carry out the tray sealing operation vary in terms of size and output speed but some core principles are in place for all designs currently available (Fig. 2.1).

The aim of this chapter is to point out the main principles of operations of the main types of machine and then look at some of the detailed designs that have an impact on the performance of the seals made by the machine. Finally, the chapter will look at the frequent causes of poor seal performance on these machines and suggest some possible corrective actions.

2.2 The principles of sealing preformed trays

The tray sealing operation involves bringing filled trays into the sealing machine in known fixed positions, covering the trays with a thin top film and then operating the sealing tools to bring a heated sealing device onto the top film (or more usually bringing the trays up to press the top film into the heating system). Heat is transferred through the top film by conduction to the junction between the top film and the preformed tray. The heat has an effect on the thermoplastic causing a seal to be made. A knife is then punched through the top film to cut out the sealed tray from the top film leaving a sealed tray and a skeleton of the top film. New top film is pulled through the machine by the scrap skeleton being wound

Handbook of Seal Integrity in the Food Industry, First Edition. Michael Dudbridge.

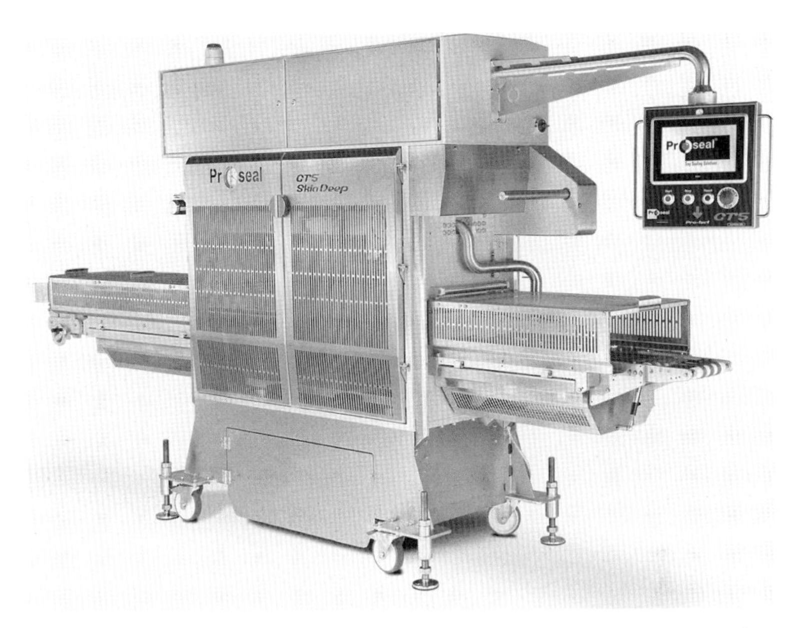

Fig. 2.1 An example of a tray sealing machine commonly used to seal foods into retail packaging. These machines, as you would expect, come in a wide variety of shapes and sizes form many different suppliers all over the world. They all have the same basic task, which is to seal a flexible top film onto a semi-rigid preformed tray. (Picture by kind permission of Proseal Ltd.)

onto the film out-feed ready for the next trays to be sealed. The sealed trays are sent out of the machine at the same time as more trays are brought into the machine for sealing.

2.3 Types of tray sealing machines

2.3.1 Small hand-operated machines

Very small machines are available that are loaded and operated by hand. These can be rotary in design or of the 'sliding drawer' type (Fig. 2.2). The same basic principles apply: the preformed tray is in a known position for the sealing tools to conduct the sealing and cutting operations on the sealing area. The output speed of these systems is dependent on the operator as a lot of the cycle time is spent loading and unloading the machine, but speeds of up to 20 packs per minute can be achieved by a skilled operator.

2.3.2 Larger, more automated machines

These have automatic tray handling to get the trays into known positions and to keep them there. The tooling for these larger machines still carries out the operation of sealing and cutting but this is often done with more than one tray at a time

Fig. 2.2 An example of a small tray sealing device which, in this case, is manually loaded and unloaded. The drawer is pushed in to make sure the action of the sealing tools takes place away from the operator.

to improve the output of the machine. These machines would typically operate at speeds of between 20 and 30 packs per minute (Fig. 2.3).

2.3.3 Very large tray sealing systems

These often operate at speeds of up to 200 trays per minute and to do this the tray and film handling systems have to be very precise. Fast machines often operate with two lanes rather than one, which doubles the complexity of the systems that control the trays and top films. The sealing tools on these machines are large and complex and as a result are more expensive and difficult to maintain at top performance.

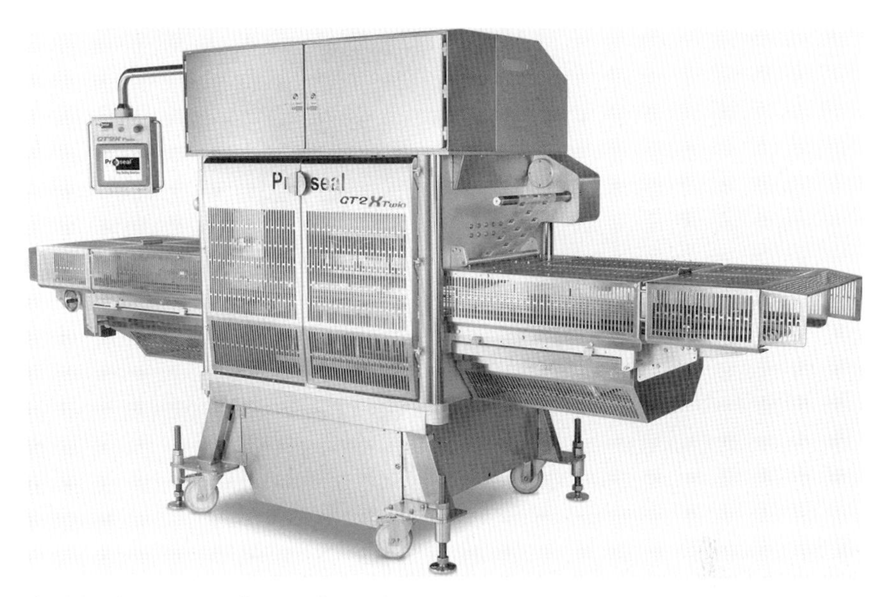

Fig. 2.3 A larger tray sealing machine where the loading and unloading of the machine has been automated to speed up the operation. Notice that the action zone of the machine is in a safety cage to ensure that no one can get their hands caught as the sealing tools come together to make the seal. (Picture by kind permission of Proseal Ltd.)

The rapid production rate of these systems means that a delay in detection of a seal integrity problem (or any other quality issue) on the packs can result in high levels of rework and waste.

2.4 The components of a tray sealing system

2.4.1 The tray handling system

Most tray sealing machines operate a system called grab and drag, which was originally patented by Mondini in Italy. Trays are marshalled onto an in-feed conveyor so that they are in known positions (Fig. 2.4). The side-to-side position is controlled using side guides to keep the trays in position laterally. The front-to-back position is controlled by using an in-feed chain conveyor made up of two chain belts. The trays ride into the machine on the chains from the filling operation upstream and when the tray is in the correct location its forward motion is stopped by a pneumatic stopper peg. The chain conveyor continues to run bringing the next tray into position while the first tray slips on the chains because it is being held back by the stopper peg. This continues until the in-feed conveyor is full and the trays are then grabbed by swinging arms that carry the trays into the part of the machine where the seals are to be made.

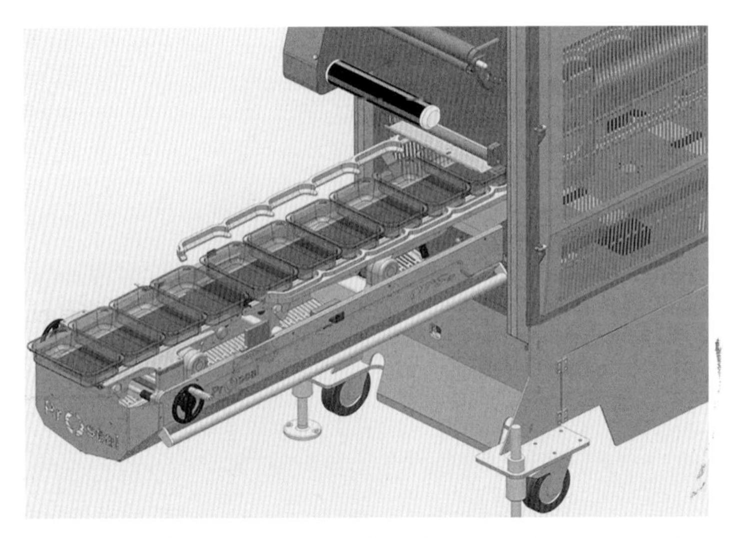

Fig. 2.4 We can see here how the trays are positioned ready to be taken to the sealing section of the machine by the grabber arms. There are many ways of positioning the packs, from using pop-up pegs to systems that use clever control of the in-feed belt systems. Here is an in-feed section of an automated tray sealing machine. The picture shows the way that the grabber arms (or transfer arms) are ready to pick up trays from the in-feed conveyor and place them into the correct position for the tooling. After the seal is made the trays are picked up again by the grabber arms and taken to the out-feed conveyor. (Picture by kind permission of Proseal Ltd.)

2.4.2 The top film handling system

Most tray sealing systems use a top film that is delivered to the machine on a reel (Fig. 2.5). The top film is unwound from the reel and pulled through the sealing area of the machine between the top and bottom sealing tools. The top film tension needs to be correct to ensure that the seals are good, so there is normally a mechanism to control the film tension. There is also a system to ensure that the top film, which is often very thin, runs in a straight line through the machine, so there are also methods to ensure that the film 'tracking' is good and always in alignment. The top film is very often printed and so there is a need for a "print registration" system to ensure that the printing aligns with the trays and the final pack looks neat.

2.4.3 The sealing tools

These are in two pieces and are of very complex and precise design. The top tool is usually heated and it is this heat that provides the seal (Fig. 2.5). The bottom tool usually has rubber inserts to ensure that when the two tools are brought together with the trays and the top film sandwiched in between there is a 'compliance fit' and the pressure that the bottom tool is exerting on the heated top tool is evenly distributed. Both parts of the tooling are designed for one particular size

Fig. 2.5 Here the top film handling system can be seen. The correct presentation of the flexible top film to the trays is essential if a good seal quality is to be achieved. If the top film creases or runs off line then leaking packs may be produced. Notice also the film storage system in the background of the picture. The correct storage of film is essential if damage is to be avoided. Even quite minor damage to the film edges will prevent it being handled correctly by the machine and as a result seal failures may occur. (Picture by kind permission of Proseal Ltd.)

and shape of tray to be sealed. The tooling is a matched pair and damage to the tooling would result if top and bottom tools from different tray shapes were used together. The tooling of a tray sealing machine makes up a substantial proportion of the overall cost of the machine. They are heavy, hot and need regular maintenance and careful handling if their sealing performance is to be maintained (Fig. 2.6).

2.4.4 The grabber arms

These devices are used to grab the trays from the in-feed conveyor and transport them to the area between the sealing tools (Fig. 2.7). Once in position, the grabber arms let go of the trays and move out of the way of the closing top and bottom tools. The grabber arms are designed to match with specific tray sizes and shapes. The use of the wrong grabber arms would result in distortion or damage to the trays and as a result the trays would be unlikely to be sealed correctly. Once the seal has been made and the tools have opened, the grabber arms come in again to grab the sealed trays. At the same time as moving the sealed trays onto the out-feed conveyor, new unsealed packs are brought into the sealing area ready for the tools to close again.

Fig. 2.6 The heart of a tray sealing machine is the tooling used to deliver the heat to the seal areas and create the required seal. The care and maintenance of the tooling is vital to good sealing performance. This includes ensuring that the tooling and other change parts (parts of the machine that are exchanged when the pack size or shape needs to be changed) are correctly cleaned and safely stored. Tooling, especially the top tooling, is very heavy and often very hot so careful and safe handling needs to be planned and training given to ensure no damage is caused to the tooling or to the operators during tooling changes. (Picture by kind permission of Proseal Ltd.)

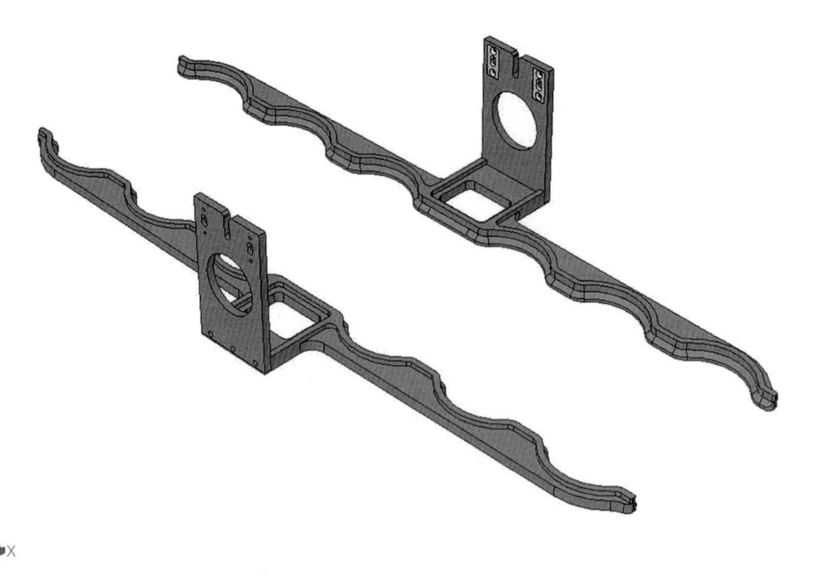

Fig. 2.7 The grabber arms (sometimes called transfer arms) have a vital role to play in the operation of an automated tray sealer. The arms can cause damage to trays (especially foil trays) if there is any misalignment or the wrong grabber arms are being used. (Picture by kind permission of Proseal Ltd.)

2.5 Tray sealing machine set-up

There are many different suppliers and designs of tray sealing machines, all of which will have different set-up requirements to make the machine ready to seal trays. The specifics for your machine will be found in the operator manual and instructions but there are some common themes that are important to all designs of tray sealer to obtain the optimal sealing performance.

As with all machines, there are very serious safety precautions that must be taken when carrying out any task on the machine. You should ensure that you fully understand the safety precautions and procedures before you carry out any such operation.

2.5.1 The top tool

This is commonly the heated part of the sealing system, so care should be taken in working with this part of the machine. The top tool also contains the devise for cutting the top film, so there is often a very sharp blade in this part. The top tool is also very heavy and should only be handled with the correct lifting equipment and protective clothing. The top tool should be regularly inspected to make sure that it is in good condition, is clean and that all moving parts are moving freely. Inside the top tool there are often springs that impart the pressure required to ensure a firm contact between the top film and the tray. These springs can sometimes become weak or even break, and as a result uneven pressure can be exerted giving poor seal performance. The cutting blade in the top tool is often overlooked but it can have a huge impact on the strength and integrity of the seal. A blunt or bent blade can have the impact of pulling a seal apart after it has been made. The timing of the operation of the cutter is such that as the tools come apart the cutter is withdrawn back up into the top tool. If the teeth on the cutter are bent they can catch the top film and pull it back off of the tray while the thermoplastic is still soft. So within half a second a perfectly adequate top seal can first be made and then pulled apart again. The top tool should be mounted correctly into the tray sealing machine and should be tightened so that it is locked into position. Some modern machines have location sensors that will alert the operator if the top tool is not correctly located. The top tool should be connected to its required services. Typically, electrical and temperature sensor connections have to be made, but sometimes compressed air and cooling water connections are also required. Any unreliable connections should be repaired (Fig. 2.8).

2.5.2 The bottom tool

This part of the tooling is usually simpler in design with no moving parts but it is still vital to the creation of a good seal on the tray (Fig. 2.9). The bottom tool is shaped to accept the tray being sealed exactly. There is often a sealing rubber that sits directly under the seal area of the tray and is the only part of the bottom tool that is in contact with the tray. This sealing rubber pushes the seal area of the tray

Fig. 2.8 The inside of a typical top tool on a tray sealing machine is a complex mixture of electronic heaters (and sometimes coolers too), sharp blades, which are often serrated, and mechanical moving parts that ensure that good contact is made between the heated seal area and the top film. All of the complexity means that a top tool is an expensive item and it needs to be well cared for if it is to perform correctly. (Picture by kind permission of Proseal Ltd.)

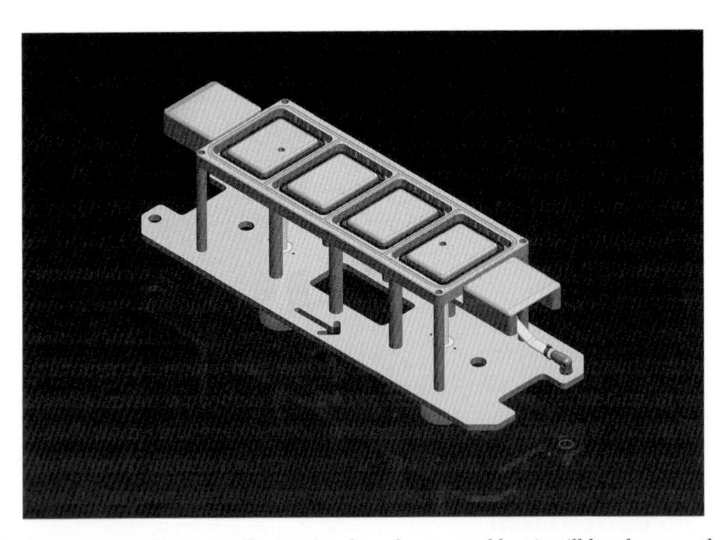

Fig. 2.9 The bottom tool is generally simpler than the top tool but it still has features that need to be maintained and cared for. The rubber parts of the tool allow the machine to cope with small variations in the packaging material but to do this correctly they have to be in good condition and of the correct elasticity. Sealing rubber can harden during use due to the impact of the constant hot/cold cycles and the effect of cleaning chemicals. (Picture by kind permission of Proseal Ltd.)

into the top film and then into the heated parts of top tool. It is designed to compensate for very small variations in tray and top film thicknesses and to ensure that the tray is evenly sealed. The sealing rubber is heated to high temperatures during the sealing operation and the constant heating and cooling can cause the sealing rubber to lose its elasticity and become harder. The effect of cleaning chemicals should also be taken into account. Powerful cleaning chemicals can also cause the rubber materials used to become harder. Once the sealing rubbers have hardened they start to split when put under pressure as the sealing tools are brought together. If sealing rubbers are hard, cracked and have pieces missing they can no longer carry out their primary function to ensure good contact between the tray, top film and the heating system and as a result the seal strength and integrity will start to fail.

2.5.3 The grabber arms for in-line machines

These components are an integral part of the sealing machine. They are made specifically for a size and shape of tray and so are often part of the changeover set required to convert the machine from one size or shape of tray to another. They need to be well maintained and correctly installed if the trays are to be conveyed correctly into and out of the machine without damage occurring to the trays or the seals (Fig. 2.10).

2.5.4 Temperature settings

The temperature of the top tool can be set on most machines via the control panel (Fig. 2.11). The temperature required to get a good seal is a function of the materials being sealed and the dwell time that the tools are together. Obviously, the shorter the dwell time the faster the machine cycle time and the higher the number of packs that can be made per minute, so the tendency is to find a dwell time that allows time for the heat from the top tool to be conducted to the sealing surfaces and use that as the starting point for setting up the sealing conditions. When using thicker top films the conduction will take longer, so the dwell time will be more than for a thinner film of the same material. The rate of conduction is also affected by the packaging materials being used. Different thermoplastics have different rates of thermal conductivity. Sufficient heat needs to be conducted to the sealing surfaces to raise their temperatures to the required level for a seal to be made, so the other factor at play here is the initial temperatures of the packaging materials. Packaging materials have a small range of temperatures where a seal can be optimised. If the operating temperature of the sealing head drifts outside of the optimum sealing temperature of the material (either too hot or too cold) then poor quality seals can be made. These can be either leaking seals or weak seals.

2.5.5 Dwell times

As mentioned above, the dwell time has an impact on the output of the machine per minute and so it is often reduced to try to speed up the packing operation. In order to make seals with a lower dwell time, the temperature needs to be raised to

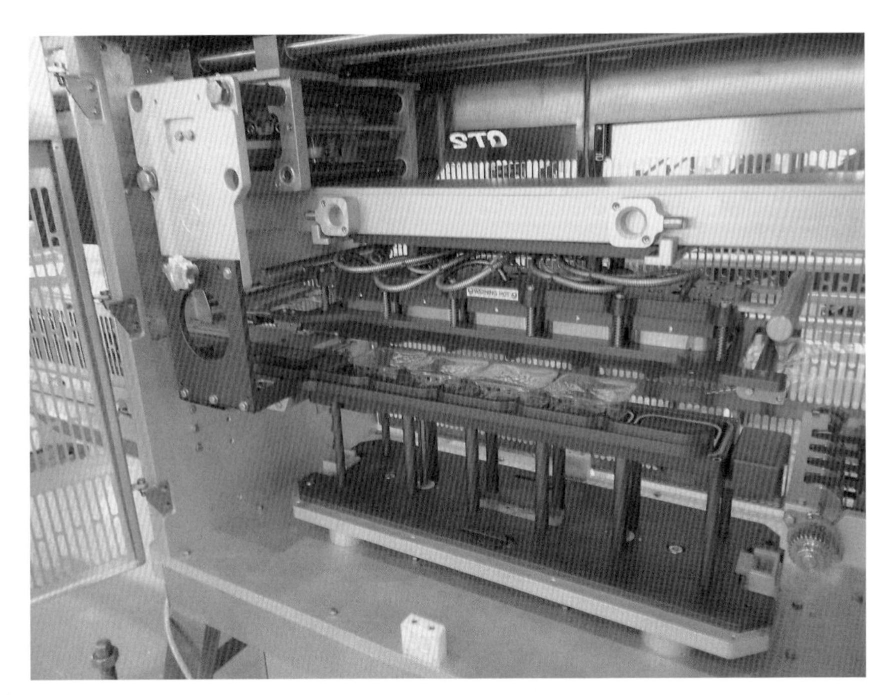

Fig. 2.10 The grabber arms have the potential to give seal integrity problems if they are not exactly matched to the tray they are handling. Damage to a grabber arm could result in misaligned trays arriving at the sealing position and also could cause physical damage to the trays. This is particularly noticeable on foil trays which retain their damaged shape if not handled correctly. (Picture by kind permission of Proseal Ltd.)

get the required energy to the sealing surfaces in a reduced time. Running machines 'hot and fast' is often seen as a good way of operating but this can often lead to other issues with the seals that are less desirable. Higher temperatures used in shorter dwell time sealing can result in the surface of the top film becoming overheated and as a result the thermoplastic can become crystallised when it cools down, giving a hard and distorted seal area. The shorter dwell time can also have an impact on the cooling time required to allow the seal to reduce in temperature and become sufficiently strong. Shorter cooling times can lead to a good seal being pulled apart before the thermoplastics have had time to cool and gain strength in the seal.

2.5.6 Top film feed

The top film is controlled by the sealing machine as it is unwound from the reel. Because the reel is very heavy and the sealing machine needs to control the film very accurately there are various tensioning devices and systems to help the machine overcome the inertia of the reel. When it is threaded onto the machine it is important that the top film is sent in the correct direction through all of the rollers and devices to ensure that it can be correctly controlled. There is often

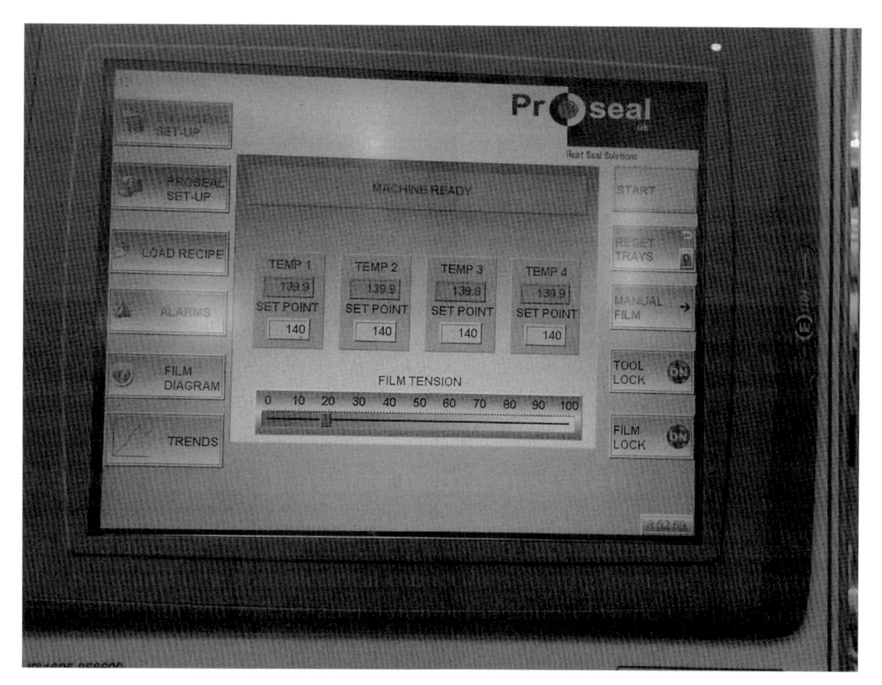

Fig. 2.11 Temperature settings are one of the major contributors to poor quality seals being formed on trays. If leaking packs are detected, the 'go to' response of many operators is to increase the sealing temperature. Many thermoplastics can be damaged structurally if they are operated at too high a temperature and so this response can actually make matters worse. Many modern tray sealing machines are controlled by different authority levels and it is for this reason that operators are unable to vary temperatures by large amounts. Temperature changes are often locked away in the control systems so that only supervisors and engineers can move the temperature more than a few degrees from the standard settings. (Picture by kind permission of Proseal Ltd.)

a threading diagram to help ensure that the machine is loaded with film in the correct way. The way in which the film is fed through the machine is one of the first things on the agenda during the training of a machine operator. The path of the film is vital to the performance of the machine, especially in terms of film alignment. If the film is not threaded through the various rollers and drives correctly the tray sealing machine will not perform well in terms of output and quality. One vital part of the film feed system is to make sure that the correct surface of the top film is brought into contact with the tray. This is obvious with a printed film but less so with a plain film.

2.5.7 Top film tension

When the top and bottom tools close it is very important that the top film is flat and does not have any creases. This is achieved by applying the correct amount of tension on the top film and also by the correct positioning of the spreading device

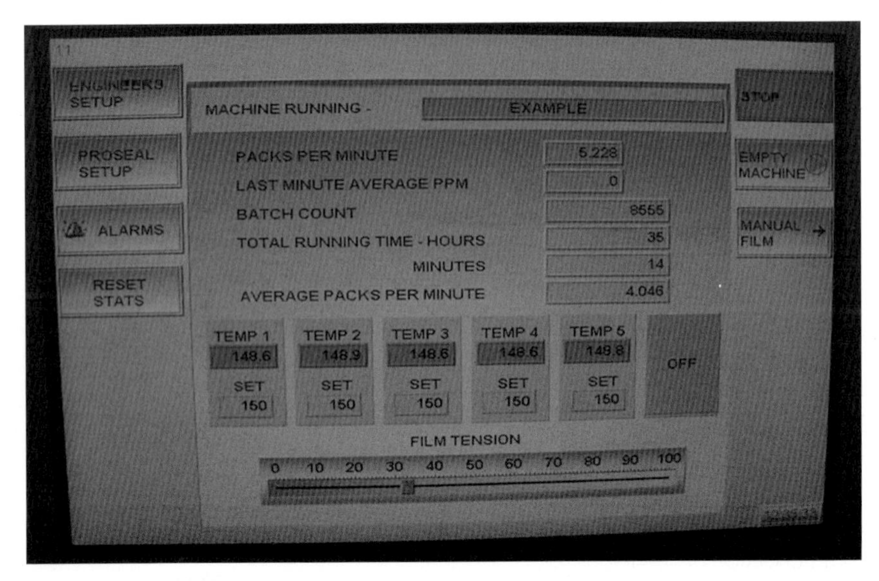

Fig. 2.12 Top film tension is adjustable and needs to be correctly set up to get maximum performance. On modern machines this is a setting on the control panel but on older machines this factor may have to be set by hand. The spreading device ensures that the film is correctly fed into the film drive rollers even though at this stage the film has holes in it caused by the seals of the previous packs. The spreading device operates just before the top film drive system and the take-up roller for the film scrap. (Picture by kind permission of Proseal Ltd.)

that ensures that the film is spread out across its full width. Any creases in the top film will result in a faulty seal on the tray being sealed (Fig. 2.12).

2.5.8 The in-feed stopper positions for in-line machines

The in-feed stoppers are devices that act on the trays while they are being fed into the machine. They ensure the correct location of the trays in relation to other parts of the sealing machine. The stopper positions can be adjusted either mechanically or electrically depending on the design of the machine. If the stopper positions are not correct the tray will not be grabbed correctly by the grabber arm and as a consequence will not be in the correct position in relation to the top and bottom tools when they close to make the seal (Fig. 2.13).

2.6 Maintenance of tray sealing machines

In order for a tray sealing machine to operate efficiently and produce consistent seals of the desired quality there are a number of routine maintenance operations that need to be carried out. There will be maintenance routines that are recommended by the machine supplier to ensure the machine operates to the expected standard and is safe for the people working on or around the machine.

Fig. 2.13 There are many ways of positioning trays on the in-feed to an automatic tray sealer. All methods have the same aim, which is to get the trays into a known position so that the grabber arms can perform their task without causing damage or dropping a tray on its way to the sealing zone. Here is a view from above the in-feed section of a single-lane tray sealer with the guards removed. In this case the spacing of the trays is controlled by sensors in collaboration with a precisely controlled in-feed belt. (Picture by kind permission of Proseal Ltd.)

The required planned preventative maintenance (PPM) of the machine will be laid out in the machine manual and this should be followed. Here we are just considering the maintenance requirements that would have an impact on the sealing performance of the machine.

2.6.1 Maintenance of the top tool

This maintenance is a mixture of hygiene and engineering functions that are needed to ensure that the top tool is in the best possible condition to make good seals (Fig. 2.14). There is a tendency for food that comes into contact with the hot parts of the tool to burn and this can lead to a build-up of carbon deposits on the heated surfaces. This carbon can build up gradually over several hours and it can be difficult to remove with the top tool in place inside the sealing machine. The food debris that can cause the problem often comes from trays that are not in the correct position when the tools of the machine close. As a result of the misplacement the trays are crushed and the food is spilled inside the machine. Recovery from these crashes is often rushed as the line needs to get going as quickly as possible to maintain the output. Debris is removed from the area, the safety doors are closed and the machine is started. Often no attempt is made to inspect and clean the heated surfaces before the food has had a chance to carbonise. Initially no problem is noticed and it is several minutes later, once the food materials have burnt onto the surfaces, that a sealing problem becomes apparent. What would

Fig. 2.14 The primary part of top-tool maintenance is cleaning and inspection. All of the parts need to be able to move freely to ensure the seals made by the tool are of optimum quality. While the cleaning is being carried out, a detailed inspection of this part of the machine should take place looking for damage and anything that is not as it should be. Carbon build-up on and around the heated surfaces is a major cause of seal integrity issues, so this should be a particular focus. (Picture by kind permission of Proseal Ltd.)

have been a few seconds to clean the top tool immediately after the crash is now at least 15 minutes of downtime as the tool is removed and the carbon is scraped off. The top tool also suffers from carbonised food in the gap between the heater and the cutting blade. If the 1–2 mm gap becomes filled with food materials this will carbonise and restrict the movement of the blade, potentially preventing the cutting of the lid film.

If the hygiene of the top tool is being well supported, the next area that needs vigilance is the inner workings of the sealing head with its moving parts, electrical components and other systems. Sealing heads are usually mounted on springs within the top tool. The springs allow small movements of the head and they allow the head to accommodate slight misalignment and packaging thickness variation as the sealing machine operates. The springs also provide the pressure that ensures that the two surfaces being sealed are pressed firmly together to facilitate the sealing process. A well maintained sealing head will be checked on a regular basis to ensure that the springs are all intact and that they are not suffering from weakening due to metal fatigue. We will see later on that sealing performance is not directly related to sealing pressure once the initial requirement of surface contact has been met (with the possible exception of sealing onto smooth-walled aluminium trays). However, the need for good contact between the surfaces is vital if good seals are to be made. The other moving part of the top tool is the cutting blade. This is usually a serrated edge blade with very sharp points that is timed to cut the top film in the correct shape and location. The film is cut after the seal is made if the pack requires an outside cut. For an inside cut, the film is cut prior to the sealing taking place. But, in both cases, the condition and sharpness of the blade can have an impact on the quality of the seal. Top tools are very heavy pieces of equipment and if they are not handled correctly the blades can become damaged. Placing the tool onto a flat surface, like a stainless steel table, can cause damage to springs and blades inside the head as well as scratch the surface of the heated parts of the tool. This can result in poor sealing performance.

2.6.2 The maintenance of the bottom tool

Bottom tools are simpler in design when compared to the top tool but are no less important in creating a good seal. A bottom tool of a tray sealing system would include a shaped support exactly matched to the tray shape and, on most machines, a compliance material to help the tools compensate for slight variations in the packaging materials. Maintenance of the bottom tool is mainly about care of the piece of equipment to ensure it is not subjected to mechanical damage. The only vulnerable part of the bottom tool is the softer compliance material. This is usually made of a grade of silicone rubber that is capable of being exposed to high temperatures. The most common faults found with this silicone rubber are as follows.

Physical damage to the rubber caused by crashes inside the machine

Plastic trays have sharp edges and these can damage the compliance layer in the event of a misaligned tray becoming crushed in the tooling. A visual inspection should be conducted when the crash is cleared to prevent a damaged sealing rubber remaining in place and causing faulty seals.

Hardening of the rubber

Over a period of time the rubber material can get harder and harder and become less compliant as a result. This is because of the constant heating and cooling to which the rubber is exposed with each sealing operation. The rubber material is pressed into the heater running at around 130–140°C (or higher with some materials) and is held there for 500 ms or so. The rubber is often distorted during this heating phase. In addition to this, once per shift (or maybe more often) the bottom tool is put through a hygiene routine where the material is exposed to cleaning chemicals. So what started as being a soft and compliant material can, within a short period of time, lose its flexibility and become hard and brittle. Gradually, cracks and fractures start to appear in the material and gaps start to open up. This results in uneven contact of the tray and lidding materials with the heated top tool during the sealing process, which in turn results in uneven seals (Fig. 2.15).

Fig. 2.15 Damaged sealing rubbers mean that the contact between the tray, the top film and the heated parts of the tool will not be even. As a result, seal failure can occur. Here you can see a sealing fault that occurred on every tray sealed with a damaged sealing rubber.

Use of incorrect materials

Different materials are sometimes used because the spare parts offered by the manufacturer of the machine are too expensive. While it is possible to source cheaper materials, every care must be taken to ensure that the parts and materials being used match exactly the requirements of the machine and its intended use.

Incorrect replacement of sealing rubber

Sealing rubber replacement is not always carried out correctly when the time comes to replace the material. It can be difficult to mount the sealing rubber into the bottom tool and if it is not done correctly the material can be stretched in some places and compressed in others. This can result in an uneven surface on the sealing rubber so that when the tray is in place some parts of the sealing area are correctly supported and some parts are either too high or too low. There is also a tendency, when the length of the rubber is being decided, to start with a long piece of the material and begin to feed it into the slot in the bottom tool where it is needed. When the outer shape of the tray has been completed the rubber is cut to length. If the rubber material is cut too short there will be a gap and if it is cut slightly too long there is a temptation to 'tuck the end in' making a bump in the sealing rubber (Fig. 2.16).

Sealing rubber shape

There has been a long debate over several years of what the best cross-sectional shape is for the sealing rubber compliance material in a bottom tool. Some are advocates for a flat top surface to match the flat area of the sealing flange on the tray. Others prefer a mushroom profile of material to try to help squeeze out seal

Fig. 2.16 The replacement of sealing rubber is a precise task that needs to be carried out with care and attention to detail. The rubber material must not be stretched or compressed as it is fitted or the top surface of the rubber will not be even. There is a particular issue on corners where it can be difficult to get the material to fit correctly. Here is an example of damaged sealing rubber that also has a large section of rubber missing. This will cause seal integrity issues on packs sealed on this base tool. (Picture by kind permission of Proseal Ltd.)

area contamination before a seal is made. The correct answer here depends on many factors – both have advantages for particular packs and both have potential drawbacks too. Safe to say that the use of a different rubber shape will have knock-on effects with sealing performance and so there will need to be other changes made to the system besides the shape of the rubber to make sure the new system is optimised (Fig. 2.17).

2.6.3 Maintenance of the in-feed, grabber and out-feed systems

This section really concerns the maintenance of the tray handling system of the machine. It is vital that the trays are in the correct position when the grabber arm attempts to pick them up. If there has been a failure in the system then the location will be wrong and as a consequence the tray will not be in the correct position or orientation when the tools close to make the seal. The resultant crash in the machine will cause downtime and potential damage to the machine. So, while tray handling is not directly involved in making the seal, a fault in this system can cause leaking packs. Routine maintenance of the stopper system and conveyor systems should look after most eventualities here.

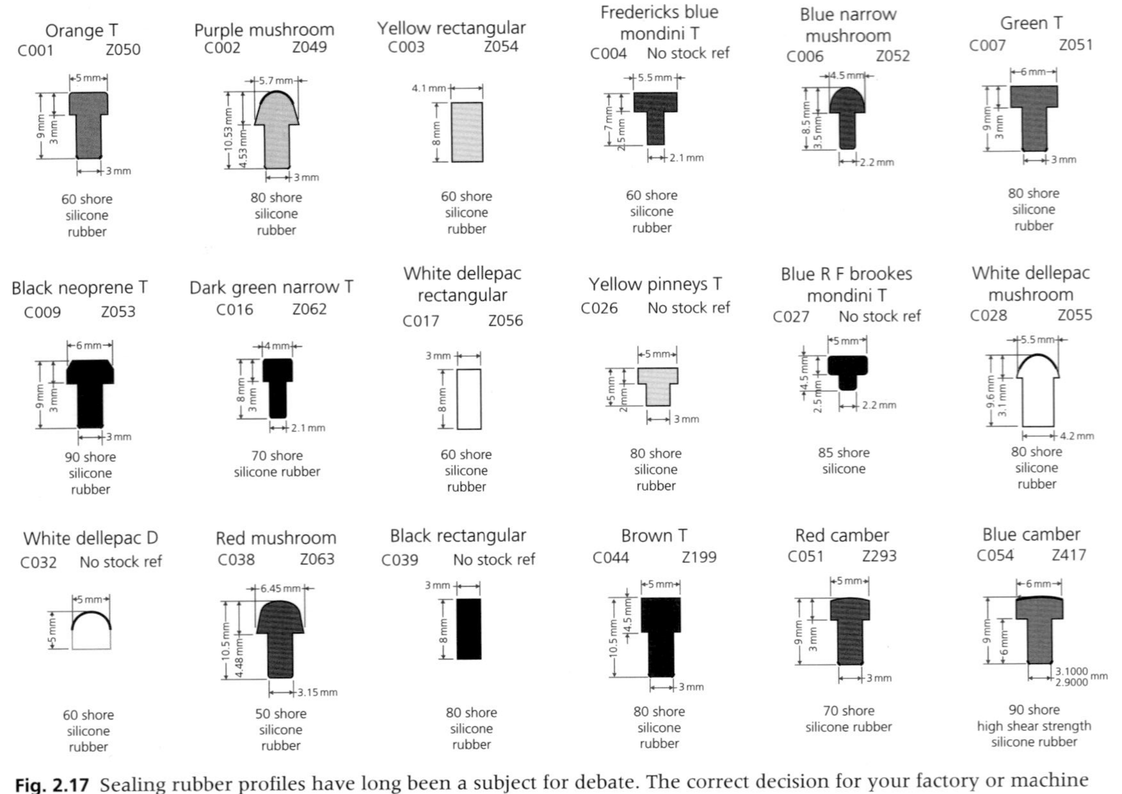

Fig. 2.17 Sealing rubber profiles have long been a subject for debate. The correct decision for your factory or machine will depend on many factors and can only really be decided following a series of tests to find the best match for your tray sealing machine, packaging type and product. (Picture by kind permission of Proseal Ltd.)

2.6.4 Maintenance of the film handling system including the scrap take-up system

The careful handling of the top film is vital to good sealing and machine performance. The route taken by the film through the machine can be quite tortuous and with each change of direction that the film takes there is a potential for the film to run off line or even break if the tension is too great. Routine maintenance of the system, which will often include hot foil date code printers and print registration systems, is as straightforward as the alignment and lubrication of rollers and drive systems. If something does go wrong and the film runs off line it could be the case that packs emerge from the machine with only half a lid film sealed in place or even that no top film is there at all. As the machine is operated it is vital that the film runs smoothly through the film clamp and via the spreader bar onto the take-up roller. To keep the film running well it is important that the scrap take-up roller system is operating neatly and the film is being pulled evenly through the machine from the in-feed reel. I have been called to a factory with tray sealing issues only to find that the clear top film had been wrongly threaded and as a result the wrong side of the film was being sealed to the tray – or at least trying to be sealed to the tray!

2.7 Machine services

Typically a tray sealing system will need a supply of electrical power, compressed air and maybe cooling water and vacuum in some cases. All of these services need to be consistent if consistent seals are to be produced. Variations in the supply of air pressure, for example, will result in air pistons operation being compromised and this could cause possible changes to the speed with which the bottom tool rises to take the trays onto the top film and then the sealing head (Fig. 2.18).

2.8 The special case of modified atmosphere tray sealing

Certain products require to be packaged in an atmosphere that is not air. Air is about 20% oxygen and 80% nitrogen and if this atmosphere can be modified it is possible to improve the shelf life of the product inside the pack or improve the appearance of the product. If a modified atmosphere is required to help protect the contents of a tray this makes the seal integrity very important; without good seal integrity the protective atmosphere will simply leak away leaving the product less protected. To get a modified atmosphere inside a seal tray is no easy task and the tray sealing machine, and in particular the top and bottom tools, have to be totally re-engineered to achieve this. This makes the tooling of the tray sealer even heavier and more complex because the open tray has to be taken inside a sealed box so that the gas exchange can occur. Trays are still transported using grabber arms and are conveyed into the correct area for sealing. The box is closed

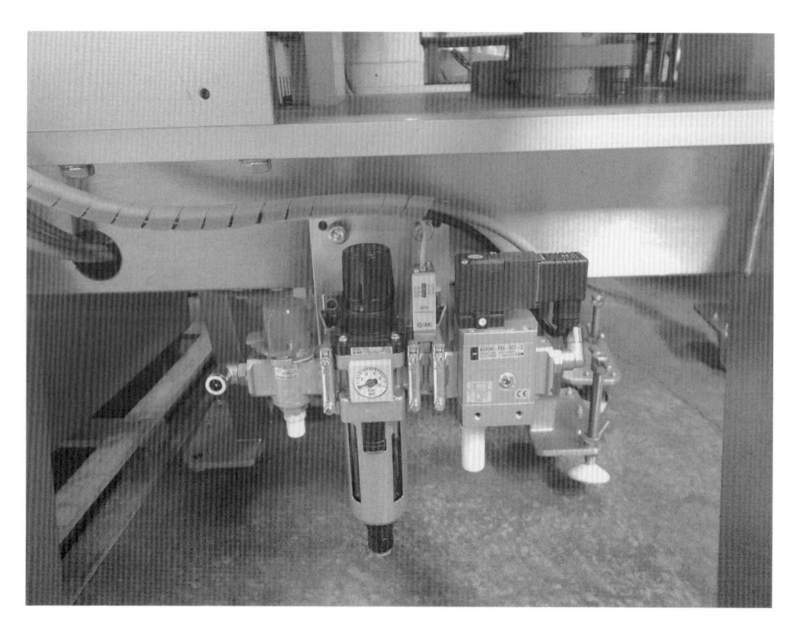

Fig. 2.18 Intermittent faults on tray sealing systems can sometime be traced back to a lack of air supply. When other machines are operating in a factory there may be insufficient air volume to drive the pneumatic system fully and at the correct speed. On one occasion in a factory I found the 'low air pressure' alarm on the control system was set wrongly so the machine was trying to operate but was unable to produce consistently good seals. The operator had been trying to improve matters by making other adjustments which then clouded the root cause of the problem. It was only when the machine was reset to standard settings that the problem became apparent. (Picture by kind permission of Proseal Ltd.)

around the trays and the normal air is exchanged for the modified gas. Once the gas exchange has occurred, the packs can be sealed. Once the packs are sealed, the box can open to allow the grabbers to take the sealed packs to the out-feed and bring unsealed packs into the box ready for sealing. The bottom tooling on a modified atmosphere tray sealing machine is much more complex than a standard set-up. The bottom tool becomes the entrance point for the modified gases and the exit point for the air being displaced. This exchange of gas all takes place in around 1 second and has to be precise, so with clever and complex design all of this can take place inside a sealed chamber.

2.9 The special case of controlled atmosphere tray sealing

This system of packaging has the intention of extending the shelf life of fresh produce products such as salads and raw vegetables. The shelf life extension is achieved by restricting access of the products to oxygen and this has the effect of slowing down the respiration rate (the cells in these products don't know they have been harvested, washed, sliced and packed and keep on respiring). It is important that the vegetables inside the packs continue to have access to some oxygen or they would die and rot very quickly. The trick is to allow them some oxygen but not too much. This access to oxygen is usually achieved using special packaging films with a known oxygen transmission rate or by making micro-perforations in the lidding film of the tray. Either way, a leaking seal would allow more than the planned amount of oxygen into the pack and as a result the respiration rate would be higher than planned and the product quality would suffer at the end of the shelf life.

So in both the above special cases the seal integrity has a direct impact on the quality of the products inside the pack and in both of these cases the packing system becomes more complex to achieve the desired result.

2.10 The trays used in tray sealing operations

Trays used to pack products are preformed and transported to the packing company for them to use. The trays are manufactured by large tray-making companies who produce the trays in very high numbers and do so to very tight specifications. Because there is a time delay between making the trays and using them there is ample opportunity to check that they have been formed correctly and that they meet the tight tolerances required for effective tray sealing using high-speed machines. The trays can be made out of a variety of materials, from thin aluminium through to cardboard materials, but the most popular types of trays are made from thermoplastics. The structure of the tray can, at first, appear quite simple but they are often made of multiple layers of materials and have complex design features to improve their performance during use. In Chapter 7 we will look more closely at the complexities of these trays but for now it's good to recognise that the size, shape and thickness of a tray can vary and that this can have an impact on packing efficiency as well as sealing performance when the trays are used. Trays are often assumed to be uniform and this is, of course, the aim of the tray-making process but the trays also have to be inexpensive and so there is an inevitable battle between cost and uniformity. Trays will have a specification that tries to quantify the variability that can be tolerated by the sealing process, and occasionally trays delivered into factories are not to specification and because of their shape or construction will cause seal integrity issues.

2.11 Top film used in tray sealing operations

The top film, or lidding film, used in tray sealing operations is more complex than it first appears. The thin film can often be made up of multiple layers of different materials to impart the desired performance characteristics. Films are often printed and often have tight specifications to ensure that they do the required job. We will look in Chapter 7 at the details of film structures and functionality, but for now it should be recognised that the lidding films have quite tight operational requirements and if the packing system is set outside of the desired levels then sealing performance will suffer. For example, as already mentioned, the 'go to' response of an inexperienced tray sealer operator to poor sealing is to increase the sealing temperatures being used. This could be the worst thing that could happen if the reason the seals are poor is that the machine is already set above the temperature window that the film/tray combination requires. Top film is supplied to the factory and should at all times meet the standard specification. The film will vary slightly but this is rarely the source of seal integrity issues, though it is unfairly blamed on occasions. Film quality is only one factor here. The age of the film as well as its storage conditions can also have an impact on film performance, so all of these factors need to be controlled to optimise the sealing performance and minimise seal integrity issues.

CHAPTER 3

The design and operation of bag-making machines

3.1 Introduction

These machines construct bags from a single roll of flat film and are sometimes known as vertical form fill seal (VFFS) machines as they firstly form a bag, then fill the bag and then seal the top of the bag (Fig. 3.1). They are typically used for the packaging of snack food products, frozen prepared vegetables and other products that need inexpensive containment.

VFFS machines can operate at high speed and are ideally suited to large-volume production.

3.2 System overview

A VFFS system is made up of the following component parts.

3.2.1 The film handling system

This takes the film from a roll of the correct width and feeds it through a series of rollers to control the film tension and make sure the position of the film is controlled and correct (Fig. 3.2). After the film is withdrawn from the roll it is formed into a tube via a carefully designed and shaped forming system. The tube of film is formed around the delivery tube through which the filling takes place later.

3.2.2 Back sealing system

As the tube is formed, the edges of the film are overlapped and a heated tool is used to seal the two edges by applying heat and a small amount of pressure (Fig. 3.3). The tube is then delivered to the bottom of the filling tube (Fig. 3.4)

Handbook of Seal Integrity in the Food Industry, First Edition. Michael Dudbridge.

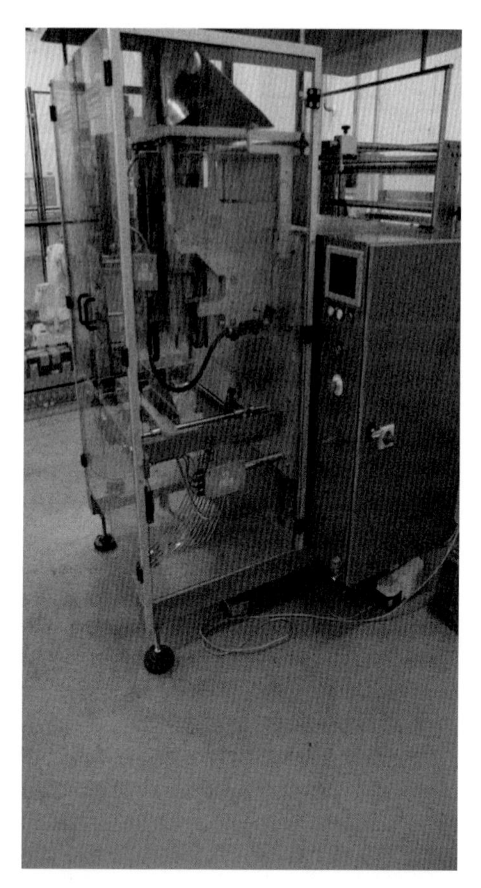

Fig. 3.1 A VFFS machine. This one is set up for large bag sizes and is yet to be connected to its in-feed and out-feed systems.

where pairs of sealing jaws form the bottom seal on the package – so now a bag has been made at the end of the continuous roll of packaging film.

3.2.3 The filling process

The next step is the filling process. Once the bag has been made and is in position, a signal is sent to the filling process to deliver the required contents to the bag. When this has been completed, a signal is sent back to the VFFS machine to carry out the next operation.

3.2.4 The top seal

The top seal is now formed and the product is sealed into the bag. At the same time as the top seal is made two other operations are carried out. The bottom seal of the next bag is made by the sealing jaws and also the filled and sealed bag is cut from the end of the tube that was formed (Fig. 3.5).

Fig. 3.2 The film handling system on high-speed packing lines is vital to good performance. Film tension is controlled here to ensure a consistent feed of correctly aligned packing material to the filling and sealing part of the machine.

Fig. 3.3 The back sealing system on a VFFS machine creates a continuous join at the edges of the packaging material and joins the flat layer of material into a continuous tube.

Fig. 3.4 The filling tube. This is the delivery mechanism that sends the product into the bag being made. The delivery tube is designed to allow product to fall by gravity down into a newly formed bag. The timing of the delivery is vital to ensure the product arrives before the top seal is made. With VFFS machines running at up to 180 packs per minute, small differences in timing can soon start to hit the overall machine performance by slowing the output or by the creation of waste product. In this example a small tube can be seen entering the much larger filling tube. This is for the introduction of modified gas into the pack – this is quite often nitrogen in snack food packaging, for instance.

Fig. 3.5 Here is an example of the sealing arrangement which forms the top seal of the bag and simultaneously forms the bottom seal of the next bag. It also cuts the newly formed bag from the end of the continuous tube of packaging material being made by the back seal section of the machine.

So the bag has been made, filled and sealed at a speed, in some businesses, of up to 180 bags per minute. In this system the performance, accuracy and timing of the machine are a very high requirement to ensure that good seals are made and that the finished package performs as required during logistics, storage and display.

3.3 Types of VFFS machine

There are many different designs and sizes of VFFS machines that are used for different reasons. Speed of operation and shape of the final package are two reasons why the designs vary, but there are also design features that allow the machine to package liquids rather than solid materials as well as features that make for good hygiene and easy cleaning if the application requires it.

3.3.1 Top and bottom sealing jaw design variations

Simple reciprocating jaws

This design makes for the overall simplest type of bag-making machine (Fig. 3.6). The pair of heated jaws operates by closing onto the tube of flexible film. As the jaws close the top seal of one bag is made at the same time as the bottom seal of

Fig. 3.6 An example of reciprocating jaw design where the jaws close, transfer their heat to the packaging material and as a result of the heat and a small amount of pressure a seal is formed. The seal bars that create the top and bottom seal simultaneously can clearly be seen along with a gap between them where the knife pushes through to cut the bag. It can also be seen that this machine has a foam rubber pad mounted below the sealing jaw. This pad squeezes the bag slightly as the jaws come together to reduce the quantity of air sealed into the bag.

the next bag. Once the jaws are closed and are gripping the film, they lower, pulling the film through the machine from the feed reel. The distance that the jaws lower is the same as the length of the bag being manufactured. Once the seal is made, a knife is triggered within the sealing jaws to separate the completed bag. As the jaws open, the completed bag falls onto the exit conveyor and the jaws return to their start position. This is a 'start/stop' machine in terms of the intermittent motion of the film through the forming, filling and sealing operations. This intermittent movement of the film can present issues in terms of the film stretching or breaking, especially if the reel of film is heavy. The inertia in the system is commonly helped with a 'dancing roller' which is used to even out the intermittent action of the system. (A dancing roller, which can be pneumatic or spring-controlled, is one that changes its position rapidly to maintain the film tension and ensure that the film is running smoothly.)

Rotary jaws

In this design of VFFS machine the heated jaws are mounted on two rotating cam systems and as the cams rotate the jaws are brought together in a pinch point. The jaws close on the flexible film and the seal is made as the heat transfers to the thermoplastic materials. Again, a knife is used to separate the bags as they are manufactured. In this system there is no tension placed on the film by the jaws to pull the film through the machine to the correct bag length. The film handling in this system is achieved with two drive belts that typically operate on the filling tube to drive the film at the correct speed, which when matched to the rotating speed of the jaws gives the correct bag length. The film in this design of VFFS machine is in continuous motion and as a result the machine can operate at higher speeds than a reciprocating jaw machine.

3.3.2 Back seal design variations

There are two basic types of back seal used on VFFS machines where a tube of film is formed as the thermoplastic is taken from the feed reel.

Overlap seal

This is where the inside surface of the film is sealed to the outside surface (Fig. 3.7). It is very common in snack food manufacture where high-speed packing on rotary jaw machines is common. This system of sealing means that there are two layers of film under the back sealing tool when the back seal is being made.

Flange or fin seal

This is a seal where the inner surface of the film is sealed to another area of the inner surface by folding the film in the forming box and bringing the surfaces together (Fig. 3.8). This system of sealing means that there are three layers of film under the back sealing tool as the back seal is made – two from the fin and one from the bag wall being pressed onto the forming tube.

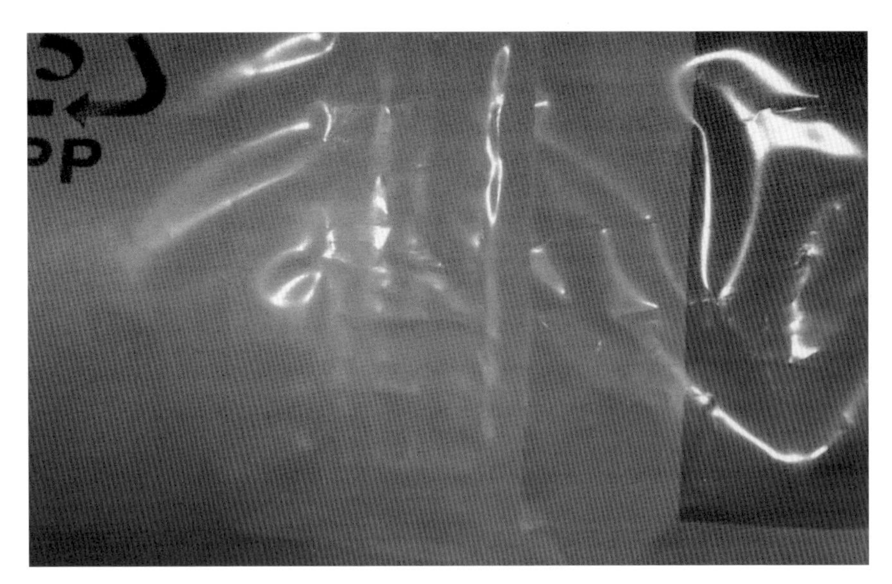

Fig. 3.7 On this type of overlap back seal the inside of the packaging material is sealed to the outside of the packaging material. This means that these surfaces have to be compatible and they are nearly always made of identical thermoplastics.

Fig. 3.8 On a flange or fin seal it is the inside surface of the packaging material that is sealed to itself. This means that the inner surface and the outer surface of the bag being made can be of different thermoplastics if required. Sometimes only the inner surface is constructed with a thermoplastic; the outer surface can even be a paper-based material.

3.3.3 The implications of the back seal for the overall seal integrity of the pack

As shown above, the two methods of producing a back seal result in two different seal structures. One has two layers of film in the seal area and the other has three layers. When it comes time to make the top and bottom seals on the package, this becomes very important. The sealing jaws that make the top and bottom seal are (hopefully) flat and straight pieces of heated metal. When they are brought into contact with the tube to make a seal there will be a potential problem. For the majority of the top and bottom seals there are just two layers of the film between the jaws. In the area of the back seal there are three or even four layers of film between the jaws. This has implications for the seal integrity as the pressure and contact between the sealing layers will not be even across the full width of the pack. This gives rise to a common fault in the seals made on VFFS systems, especially on machines running with thinner and harder films. The multilayer of film in the back seal area holds off the sealing jaws and as a result small leaks occur immediately next to the back seal where the packaging thickness changes from four (or three) to two layers (Fig. 3.9).

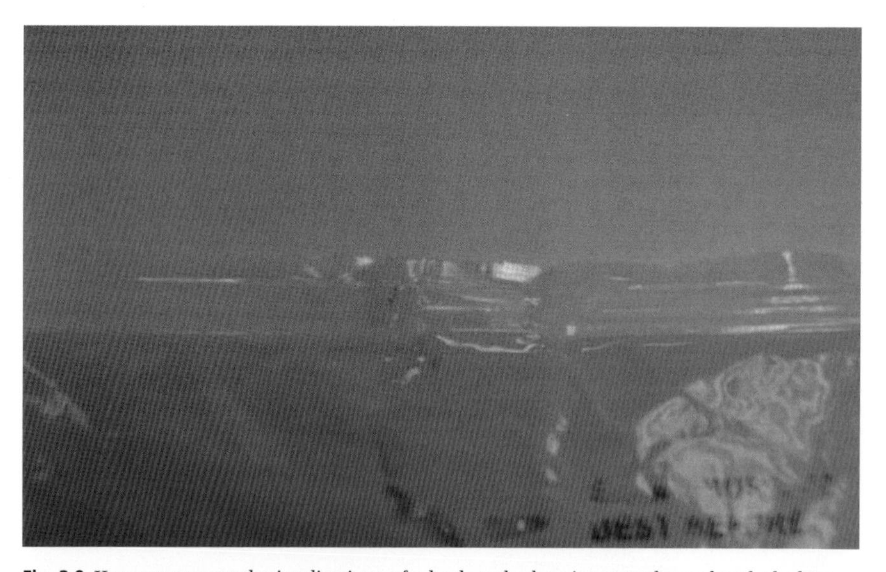

Fig. 3.9 Here we can see the implications of a back seal when it meets the end seal of a bag. There is opportunity for the bag to leak because there is a small channel at the point where two layers of packaging material change to three or even four layers of material. (Picture by kind permission of RDM Test Equipment Ltd.)

3.4 Components of a VFFS machine

As with nearly all packaging machine systems the design of the machine can be broken down into smaller sections.

3.4.1 Film handling and forming system

This system is designed to deliver film to the machine in a way such that its behaviour is predictable. At its simplest it is a way of taking film from a reel and guiding it in a controlled way to a former that converts the flat film into a continuous tube. On the way to the former the film may be required to have information printed on it, such as a date code, and may be required to be exactly positioned in order to get the correct print registration. If the film on the handling system loses tracking and moves to one side (even fractionally) its feeding into the former will be compromised and the back seal could become a problem if the materials are not adequately overlapped or aligned (Fig. 3.10). This is especially a problem if the width of the film has been optimised to reduce the overlap to save on packaging materials. The margin for error has been reduced and this can lead to more back seal failures if the film handling is underperforming. The final element to be aware of in any system handling plastic films at high speeds is the build-up of static electricity that can occur. This can make thin films change their behaviour and as a result can have an impact on sealing quality and performance.

Fig. 3.10 The film on this VFFS machine has been misaligned and you can see that the back seal is not being formed correctly. It looks untidy and will, almost certainly, leak.

3.4.2 Multiple functions of the forming system

The forming system on a VFFS bag maker is a part that has multiple functions. It is precisely engineered to form the required tube of film and has a very important role to play in guiding the film into the correct position. The forming system on a VFFS machine is a key component of the change parts of the machine that allow it to manufacture different widths of bags. As a result of this change-part role the forming system may spend some time away from the machine while a different size pack is being made. The care of the forming system is paramount if the performance of the machine is to be protected. Damage to the forming system (no matter how small) can have an impact on its function and as a result the performance of the bags being made can be affected. The forming system is a piece of carefully designed three-dimensional geometry. Slight changes in shape or roughness of the surface can cause downtime, waste and poor sealing (Fig. 3.11 and Fig. 3.12).

One other function of the forming system is to act as the entrance to the bag for the product feed system. As a result of this role, the forming system/feed tube may need to be cleaned on a regular basis to prevent a build-up of product or bacteria in the feed system. Hygiene systems for this vital part of the machine need to reflect its important role and it should be treated with care and be stored correctly whenever it is not mounted on the machine. Factories where forming systems are abused are likely to suffer poor machine performance.

Fig. 3.11 Here is a well-maintained forming system that is doing a great job of converting the flat packaging film into a three-dimensional tube. Here the packaging film is transparent but no creases or folds can be seen. Compare this with Fig. 3.12.

Fig. 3.12 Here is a forming system that is not doing a very good job. You can see folds and creases in the film, which indicates that the film is not flowing smoothly over the former to create a tube from a flat sheet of packaging film. In this case the film is metallised.

The forming/feed system also provides the platen against which the back sealing tool forms the back seal on the pack (Fig. 3.13). This area of the filling tube can cause problems because of a build-up of thermoplastic materials on the tube. This is sometimes overcome by the use of polytetrafluoroethylene (PTFE) tape. This tape is heat resistant and has the bonus of a very slippery surface to reduce the deposits of thermoplastics in that area of the machine. The downside of PTFE tape is that it can become damaged and start to present an uneven surface for the formation of the seals. This uneven surface, if it goes unnoticed, will provide uneven pressure as the back seal is made, and as a result the machine could start producing weak or leaking seals.

To try to overcome this issue, some machines are designed with the ability to heat seal twice on the back seal. This is especially the case where the bags being made are a lot shorter than the filling tube. The back sealing bar is usually designed to be one and a half times the length of the longest bags to be made but this is not usually a change part on the machine so when shorter bags are being made it is possible for each part of the back seal to be heated and pressed twice.

3.4.3 The sealing jaws

These are the heated devices that make the top and bottom seals on the bag. The surface of the sealing jaws can have a big impact on the seal strength and integrity. We have already considered the issues that occur in the area of the back seal on a VFFS system, but there are other considerations too in this vital area of the package. There are a great number of different surface profiles used in VFFS machines. The profile chosen will often depend on the product being packaged as well as the packaging materials being used.

Fig. 3.13 Here we can see the area on the outside of the filling tube that is covered with PTFE tape. It is this area that is used as the platen for the back sealing tool to press against to form the back seal of the bag.

Horizontal pattern

This pattern is often used in the snack food industry where there can be an issue with crisps and crumbs getting trapped in the seal area as the top and bottom seals are made. The theory is that the horizontal pattern allows at least one of the three or four 'seals' to be good. The other aim of this sealing pattern is to produce a seal that is peeled evenly without too much drama. Because of the relatively light weight of the product inside the bag the seal need not be overly strong, but good seal integrity is needed as snack foods are often packed in a protective atmosphere of nitrogen to reduce the possibility of oxidation of the fats inside the crisps and so extend the shelf life (Fig. 3.14).

Vertical pattern

This type of sealing pattern is often used in confectionery packaging and aids opening by allowing the packaging material to tear from the top of the pack (Fig. 3.15). To take off a corner in this way is a good easy-open feature but has the disadvantage of being susceptible to channel leaks where the back seal meets the top and bottom seals on the pack.

Fig. 3.14 Here is a typical horizontal seal pattern used on packages where seal integrity is important. The horizontal pattern allows the thermoplastic material to flow into continuous lines across the pack and this gives multiple opportunities for a complete seal to be made.

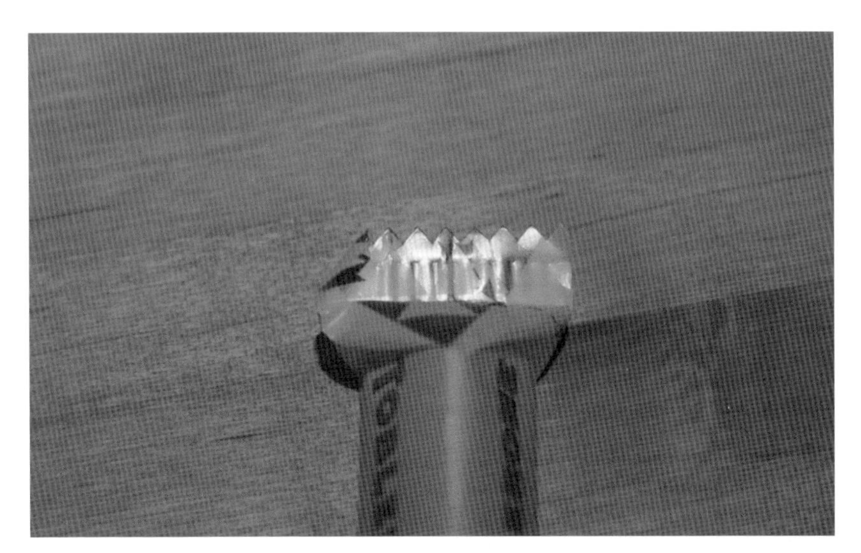

Fig. 3.15 A typical vertical seal jaw pattern often seen on confectionery packaging where a small leak is not too big an issue. The vertical pattern combined with a serrated cutting knife gives an easy-open feature to these packs.

Both the vertical and the horizontal jaw pattern require the precise pairing and positioning of the jaws to ensure that the required contact and heat transfer occur when sealing. If the patterns are not matched it is possible to have an area within the seal where the seal is weak or not formed at all. One of the functions of the sealing pattern is to give areas within the seal where thermoplastic material can flow under the influence of the sealing heat and pressure. To create a good seal the thermoplastic layer in the packaging film must be thick enough to create enough semi-fluid material to flow and fill all of the voids in the seal. If the thermoplastic layer is too thin or the jaw pattern is too bold there is a danger that insufficient material will move to fill the voids and the seal will be weak or leak.

Flat seal jaws

This type of jaw pattern is used on packaging systems where the materials are thicker than usual and where the limiting factor in the speed of the sealing process is the rate of heat transfer and conduction to the sealing layer thermoplastics (Fig. 3.16). In order to optimise the heat transfer the maximum contact is required between the hot metal sealing jaws and the materials, and this is best achieved using a flat sealing jaw with no pattern. It must be emphasised that flat sealing jaws and large seal dwell times can result in large amounts of flow of thermoplastic material from the seal area. Some of this flow will be to the part of the pack where the product is, so care should be taken to prevent product contamination by ensuring that the sealing layers are adequate but not too thick.

Milled sealing pattern

Sometimes called a hatched pattern, the milled sealing pattern is a very fine pattern in the surface of the sealing jaws (Fig. 3.17). There are some claims that the pattern increases the rate of heat transfer to the packaging material and also can deform the surface layers of the packaging film if the materials are able to be

Fig. 3.16 Here is a flat and highly polished sealing jaw. The need to keep this design of jaw very clean is vital as any debris build-up will cause the jaw to be 'held off' its correct position and as a result it is unlikely that the seal will be adequate.

Fig. 3.17 A milled pattern is often used in the sealing of powdered foods. Here the pattern has been used on a packet of dried soup mix. In this case the milled pattern has been selected due to the dusty environment in which this item is packaged.

changed in that way. For certain applications both of these features are desirable, for instance to allow faster packing speeds of thick packaging materials. Pet food pouches often used a milled sealing pattern to overcome the issues of the thick packaging materials. There are also claims that the milled surface reduces any build-up on the surface of the jaws by carbonising and breaking up the contamination material before it becomes a major problem.

3.4.4 Support arms

VFFS machines are often used to pack heavy items, and special adaptations can be made to help the machine perform better. When the sealing jaws have done their work and a sealed bag is released from the machine, the sealing jaws retract and wait for the next products to come down the filling tube inside the pack. The bottom seal of the next bag will still be hot as the jaws retract and the next products fall. If the seal has not cooled sufficiently the thermoplastic materials may still be mobile and as a result the weight of the next product hitting the bottom seal can force it open. The opening can either be as a result of the impact – the products can fall from some height into the bag and have a lot of momentum when they hit the bottom seal and stop – or it can occur by creep as the weight of the product rests on the hot seal area. This problem is sometimes overcome by the use of support arms. These stay in place just above the bottom seal when the sealing jaws retract. They prevent impact onto the bottom seal and also allow time for the seal to cool before the weight of the product is released onto it. The design of the support arms and the timing of their operation vary from application to application but the principle of operation remains the same.

3.4.5 Exit conveyor systems

When a bag is made and it is cut from the end of the film tube it usually falls by gravity onto the exit conveyor. Designs in this area vary depending on the product being packed but it is usually a parabolic system that allows the filled bag to move quickly away from the jaws and fall vertically downwards (Fig. 3.18). The momentum of the bag is then used to move it onto the exit conveyor system by sliding on a curved guide. The designs in this area are very important in high-speed packing operations where it is vital that the bags are moved efficiently to the exit of the machine to separate them sufficiently for the next section of the line, which is usually an automatic check weighing system. It is also vital that the packs be treated as gently as possible at this stage. The seals are still hot and if too much force is put onto them they will fail. The parabolic exit chute is seen as the best way to guide the filled bags to the exit of the machine, especially on systems that are packing heavy items such as potatoes, where the weight of the bag contents has a great potential for putting excessive forces onto the seals before they have had time to cool and gain their optimal strength.

3.4.6 Controls of a VFFS system

A VFFS machine is usually fed with product from above using an automatic weighing or dosing system. The correct weight or volume of product is created and sent down the filling tube to be sealed into a bag. The timing and coordination of the weigher/doser and the VFFS machine is vital – especially on a high-speed production line. For example, on a high-speed system packing snack foods into

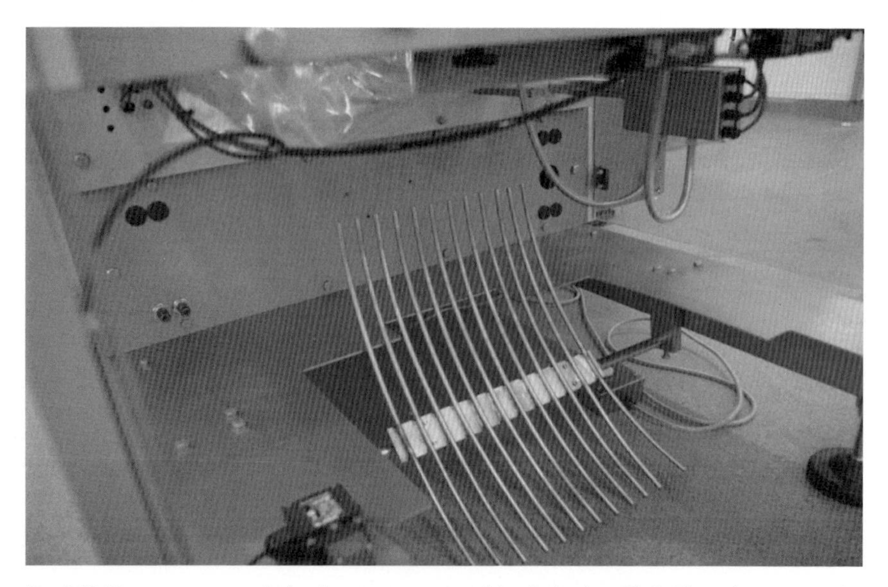

Fig. 3.18 Here we see a typical exit system on a machine being installed. The exit conveyor is not yet in place but you can clearly see the important parabolic shape to the exit chute.

bags it is sometimes the case that the system of weigher and bagger is running at speeds of up to 180 bags per minute. That equates to three bags per second. Running at this speed there will be up to four portions of the snack food on their way to the bagger at any one moment. The weigher releases the snack food which then falls by gravity into the filling tube. This can often be a fall of 2 m or more and this can take over 1 second to complete. Because of this there is a need to know the 'flight time' for the product in order to coordinate the weigher and bagger correctly. The jaws on the VFFS system have to do their job where there is a gap between the products that are in flight to ensure that no product ends up being trapped in the sealing area. So the controls for the timing and coordination of a VFFS machine will have the ability to bring forward or delay the point at which the jaws close to enable it to do its job. The other controls on the VFFS system will include time and temperature settings for the back seal and also the top and bottom sealing jaws.

3.5 VFFS machine set-up

A VFFS bag-making system requires very careful set-up to ensure that it runs efficiently and produces good quality packs that are correctly sealed.

It is key to the set-up of a VFFS system that the correct dimensions of forming system and filling tube are used as these are matched to the width of the plastic film being used to make the bags. When formed into the required tube, the width of the film needs to be the same as the external circumference of the filling tube plus the allowance for the film overlaps required in the back seal. If the filling tube is too small for the width of the film there will be too much material in the back seal and as a result the pack will not look neat and will not run well on the machine. I have seen products where the wrong filling tube has been used for the width of the film and the information on the back of the bag has been covered over by an enormous back seal. If the filling tube is too large for the width of the film it is likely that the bag seal will be too narrow and that while the VFFS machine is operating the back seal will move apart and the bags will not be sealed.

Film feeding on a VFFS machine is very important. It the film is not fed correctly the tension in the film will not be correct and as a result the film will wander from side to side as the machine operates. This will cause failures in the back seal and also faults such as date codes not being in the correct location or the pack length becoming variable because the eye mark on printed film has wandered away from the sensor used to detect it.

Temperature and dwell times are important on any sealing machine and a VFFS system is no different. The required settings will depend on the packaging material type and thickness and these settings will also vary with the temperature of the room and hence the temperature of the film arriving at the sealing jaws. Uniquely for the VFFA type of sealing system, the product being packaged rides down the filling tube prior to the package being sealed. If the product is frozen this will mean that the filling tube will be cold and this will have an impact on the film temperature and hence the seal temperatures and dwell times chosen to achieve a good seal.

VFFS systems can produce complex shapes of bags by cleverly tucking in the sides of the package before the top and bottom seals are made. These gusseted bags can give a different and pleasing result in a retail environment but can also present additional seal integrity issues if the machine set-up is not correct. The create a gusset on a bag means that the top and bottom seals now have to seal through more layers of materials and there are more positions where potential leaks in the packaging can occur.

3.6 Maintenance of VFFS systems

General maintenance of a VFFS machine is vital to continued good performance and good seal quality. If maintenance is not carried out or is inadequate then the performance of the VFFS, like any other machine, will suffer. There are some particular areas of a VFFS machine that need careful attention as issues in these areas have a large impact that can go unnoticed for long periods if not corrected.

3.7 The feed of product into the VFFS

The feed of product to a VFFS is often from an automatic weighing system. We have seen in the set-up section that the links and set-up of both systems are vital for good performance. On a fast production line the relative timing of the weighing with the movement of the sealing jaws is a vital factor of success, so anything that interferes with the timing is likely to cause a major problem in sealing the product inside the bag. A slow-moving pneumatic cylinder on an automatic weighing machine hopper will mean that the product is dropping a few milliseconds later than it should. This can be enough to throw the timing of the VFFS machine out and cause the product to be caught in the seal. This might only be happening on one out of 16 hoppers on the weighing machine, so not every pack will be faulty – maybe 1 in 4 or 1 in 5 will show the product-in-seal fault.

Sticky product being packed will also cause the same timing issue. Sticky does not necessarily mean a product covered in syrup. Slightly defrosting frozen vegetable can stick to the walls of the hoppers on automatic weighing machines and as a result the decent of the product into the VFFS can be delayed by a few milliseconds. This can be enough to cause top and bottom seal faults, with the product getting trapped in the seal. The same applies to wet washed leaf products. Slightly too much water on a lettuce leaf can cause its decent to be delayed, with the obvious consequences.

So it can be seen that sealing problems on a VFFS might have the root causes further upstream and the maintenance issue may be on the lettuce leaf drying system and not on the VFFS machine.

3.8 Sealing jaw maintenance

The sealing jaws on a VFFS machine can be operated in several ways: mechanically, electrically or using compressed air. Issues with compressed air systems have, in the past, occurred because the air pressure has been insufficient to produce the required motion of the jaws. This would usually be picked up by an air pressure sensor and the operator would be alerted, but this is not always the case. The air pressure sensor is often on the air inlet to the machine and all can be well at that point. If an air leak or a kink in an air feed tube has occurred inside a machine this can result in the incomplete closure of the sealing jaws and as a result faulty seals will be made (Fig. 3.19).

Unbalanced sealing jaw movement can have several root causes. The bearing on which the jaws move can be worn or not well lubricated and as a result one side of the jaws is fractionally held off making its final position. The result of this is uneven seal pressure across the jaws, which could result in weak seals. The same effect can be seen if the heated surfaces of the jaws become dirty (Fig. 3.20). The carbonisation of the contamination can cause a very hard deposit to build up on sections of the jaws. This will prevent the jaws from closing correctly and as a result faulty top and bottom seals can be made. This build-up needs to be removed using some safe procedures. The jaws will be hot if this task is being carried out

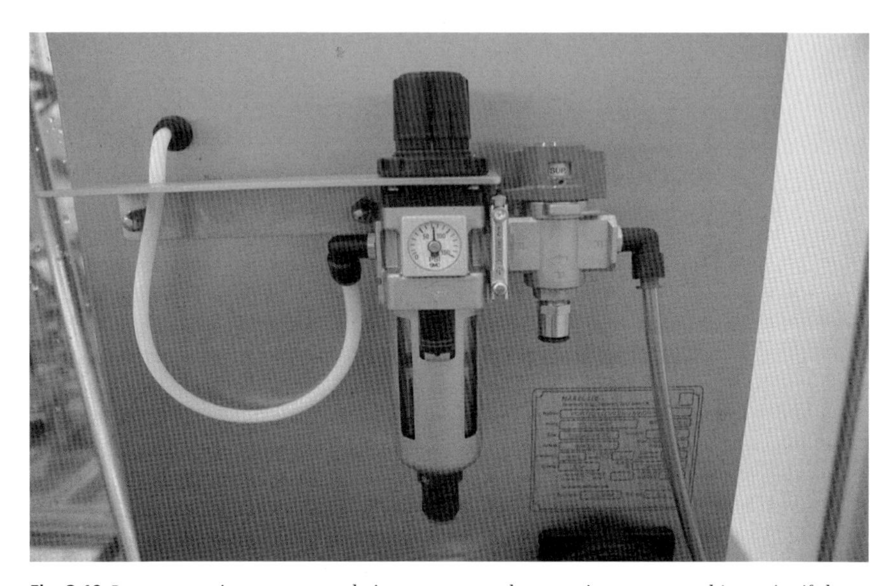

Fig. 3.19 Low or varying compressed air pressure can have an impact on seal integrity if the movement of the VFFS machine is altered or delayed. Correct air flow and air pressure are vital to the smooth operation of the machine. I have investigated several intermittent faults on seals that have ultimately been tracked back to variations in compressed air services to the machine. These were overcome by fitting a local buffer tank to smooth out the variations and keep the machine running correctly.

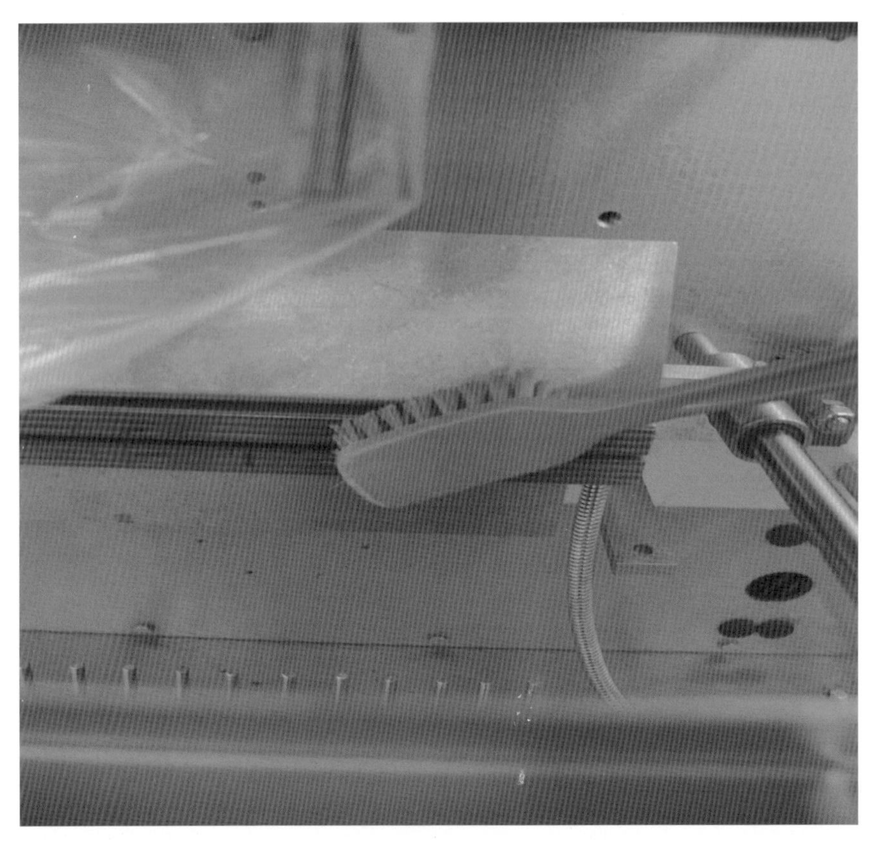

Fig. 3.20 Keeping the sealing jaws clean is a routine requirement in most packing operations. It is especially important in dusty environments where dust that has landed on the jaws and carbonised needs to be removed on a regular basis to prevent seal integrity issues. Even a small piece of carbonised food can be enough to hold off the jaws as they close, resulting in a faulty seal being made. Build-up of inks and of thermoplastic material can also occur, especially if the VFFS machine is running at high temperatures using printed films.

during a production run, so hand protection must be worn. The jaws would usually be cleaned using a wire brush to get the carbonised material away. These small pieces of carbonisation can flick off the jaw, so eye protection should be worn too.

Some sealing jaw surface patterns need to be correctly aligned to work properly (Fig. 3.21). A peak on one jaw is required to mesh with a valley on the other jaw. This maximises the heated surface of the packaging material resulting in the thermoplastic layer activating more quickly. If a peak hits a peak then there is very little surface area through which to transfer the heat to the thermoplastic. The other issue with jaw alignment is the need to ensure that the jaws are operating in parallel to each other and are correctly positioned to carry out their operation effectively.

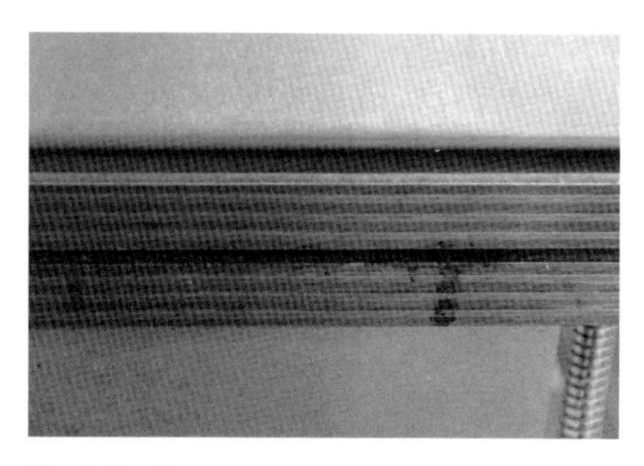

Fig. 3.21 This is a comparison with a matched pair of sealing jaws. The high points on one jaw correspond to a low point on the other jaw. This gives good performance providing the jaws are correctly aligned. Slight misalignment or wear can cause sealing performance to deteriorate.

3.9 Jaw and back sealer heating systems

On VFFS machines the heat for sealing is usually supplied using a cartridge heater. This is a sealed tube with electrical connections coming from one end. Cartridges have a life span and will fail after a number of hours of work. Cartridges come in a variety of power ratings, lengths and diameters and in the dynamic environment of a packing operation the temptation to use a different specification of heater is always there when the machine is needed to run. It could be that the last one of the correct heaters was used up last week and the only one we have looks the same, 'so that will be OK, won't it?' It could be that 'this new supplier of heaters is offering them at a much better price and the maintenance budget is under pressure so we need to make savings'. There are lots of reasons why this change of specification is a good idea but changes in this area should be done very carefully to ensure the substitute heater cartridge is not going to cause issues with machine performance (Fig. 3.22 and Fig. 3.23).

3.10 Film forming system

On the face of it, the guide system that forms the flat packaging film into a tube has a simple function which it carries out with no moving parts. The forming systems for VFFS machines are complex pieces of three-dimensional design that need to be maintained just like any other part of the machine. The fact that the forming system is a change part to allow different sizes of bags to be made means that it spends a proportion of its life not mounted in the machine but waiting for the next size change on the machine. This is where the carefully designed shape can be altered accidentally so that the film does not run as well as it should. The film forming system also contains the filling tube so it can become contaminated with product and requires cleaning. This is where more damage can be done to render the part in need of maintenance. The surface should be smooth to allow

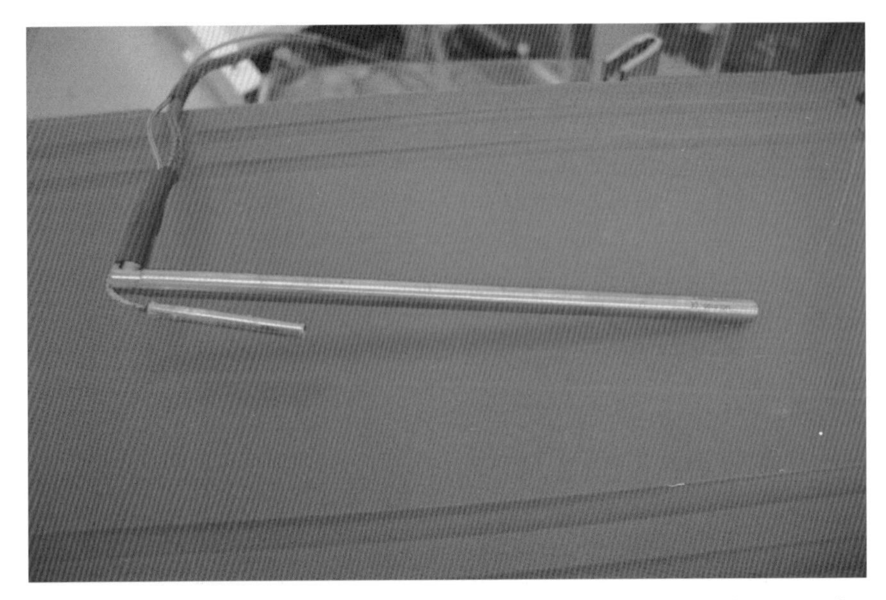

Fig. 3.22 Here is a cartridge heater that has been removed from a sealing jaw. The picture also shows the temperature sensor and it can be seen that on this machine the sensor is much shorter than the cartridge heater it is controlling and is positioned at one end. It is also common for less expensive cartridge heaters to have the majority of their heat generated away from the electrical connections to try to lengthen the life of the heater. So it is quite possible that the temperature could vary across the width of the sealing jaw.

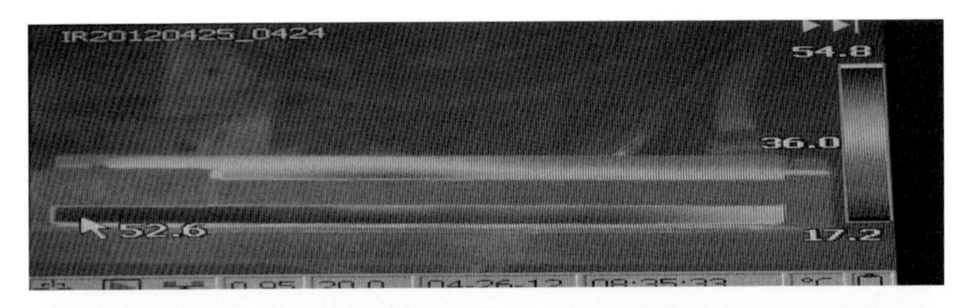

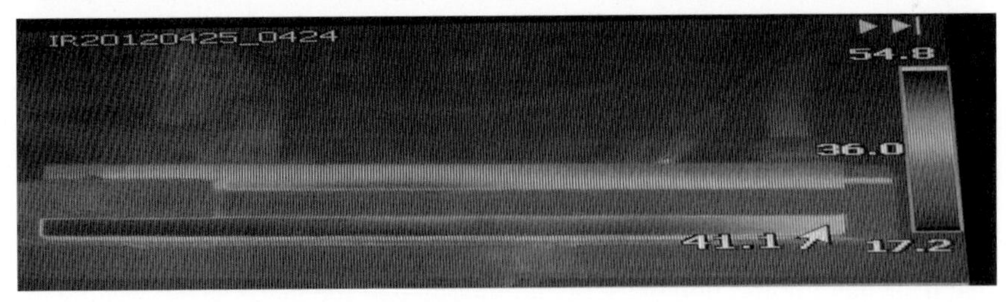

Fig. 3.23 Here is a thermal image of a cartridge heater being operated outside of its sealing jaw. The temperatures were operated below normal for safety reasons and the heater was allowed to equilibrate before this thermal image was recorded. It can be seen that the temperature of the heater varied by over 10°C, from 41.1 on the right of the heater to 52.6 on the left. This was a heater cartridge that was 20% less expensive than the standard replacement part and, yes, the electrical connections on this heater were at the cooler right-hand end. The seal integrity issues on this machine were tracked back to this variation in sealing temperatures.

the free running of the film and also should be free of scratches and dents that could cause the film to tear or run off line.

3.11 Calibration

As with all pieces of electrical sensing equipment, the sensors on a VFFS need to be calibrated to make sure that consistent messages are sent to the control systems (Fig. 3.24). One area that is of particular importance is the calibration of the temperature sensors on the sealing jaws and the back sealer. The temperature sensors for the thermostats that control the jaw temperatures are usually mounted in one position in each jaw. The sensor is not at the jaw sealing face but is usually mounted in the same space as the heater cartridge. You would expect the signal from the sensor to be different from the temperature of the jaw surface, and that is acceptable. The important point here is that the sensor needs to be calibrated to ensure that consistent seal temperatures are achieved over a long period of time. For example, a calibrated sensor on a seal jaw is indicating on the VFFS control panel that the temperature of the jaw is 150°C. The jaw surface temperature was measured with a surface probe and found to be 135°C. That is acceptable as long

Fig. 3.24 Calibration of temperature sensors is an important control to have in place for the long-term capability of the sealing system. Here a surface probe is being used to check the surface temperature of the back sealing jaw so that it can be compared with the indicated temperature on the control panel of the VFFS. The actual temperature of 104.8°C was achieved when the controls were set to achieve 130°C. An investigation and a recalibration were carried out.

as the next check reveals a consistent difference. At the next jaw temperature check it was found to be 130°C but the control panel was still reading 150°C. There has been a change in the calibration of the sensor and this needs to be rectified to ensure consistency in the system. I have visited factories that have indicated temperatures of 240°C on their control panel with a sealing jaw temperature of 130°C. That's all well and good until the sensor breaks and a new one with a different calibration has to be fitted. The operators would set the temperature at 240°C as usual and might get a sealing jaw temperature of 220°C rather than the desired 130°C. The point is they would not know where to set the machine to get the desired sealing temperature at the jaws. That's going to be a stressful start to the shift and presents the need to communicate the new required setting to all the machine operators.

3.12 The special case of modified atmosphere packing with VFFS

Many products packed using VFFS systems are packaged in a protective atmosphere to extend the shelf life of the product. For example, many snack foods that are high in fat are packaged in an atmosphere that is low in oxygen to reduce the oxidative changes that would alter the flavours and might turn the fat in the product rancid. To achieve this, the product needs to be surrounded by nitrogen when the top seal is being made on the bag. This is normally achieved by feeding the gas into the filling tube at a rate that is slightly greater than the volume of gas that is being sealed into bags each minute (Fig. 3.25). So if each bag contains 20 g of crisps and 0.25 l of gas and the VFFS machine is running at 160 bags per minute, the nitrogen feed has to be at the rate of 160 × 0.25 = 40 l per minute plus 5% to provide a total of 42 l per minute to ensure that each bag is flushed with nitrogen.

Depending on the type of product being packed, this method will not exclude all oxygen from the pack. Normal air is made up of about 80% nitrogen and 20% oxygen with a small amount of other gasses. Any air that is captured inside a product will be carried into the pack. It is only the bulk of the air that is displaced by this method of gas flushing, but for most purposes the oxygen level in the final pack will be acceptably low at around 1%. The normal air is said to be 'entrained' in the product and is carried into the pack through the surrounding nitrogen atmosphere. There is one possible problem with this method of packaging product in a protective atmosphere and it is to do with the flow of gasses as the product falls into the bag.

Consider this in stages. First, a sealed end of a continuous tube is formed by the back seal heater system and the bottom seal heater system. This sealed end of the bag is then filled with nitrogen gas at the correct rate. The product is then released by the weighing system and it falls into the bag that is already full of nitrogen gas. As the product enters the end of the sealed tube of packaging

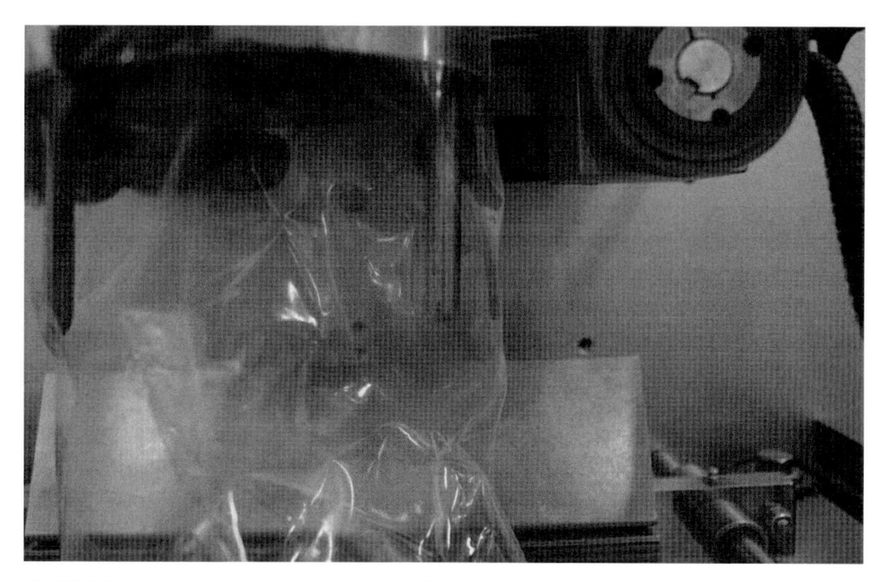

Fig. 3.25 Here we can see the tube that is feeding nitrogen gas into the bags as they are sealed. This machine looks a bit mucky – it had just finished an 8-hour shift!

material it will displace its volume of gas from the bag. So, just before the top seal is made on the bag, a volume of nitrogen the equivalent of the volume of the product in the bag is displaced and flows upwards and out of the bag that is about to be sealed. This flow of gas upwards is in the opposite direction to the next lot of product flowing down the filling tube and into its bag. So, on high-speed VFFS packing systems it is an important consideration that the product falling by gravity will be resisted by the nitrogen flowing in the opposite direction. This has the consequence of marginally extending the 'flight time' of the product from the weigher to the bag and this small extension in time is enough, when a machine is being run at maximum speed, to cause product to be trapped in the seal area. This effect can be minimised by controlling the flow of nitrogen into the system to make sure that it is just enough to keep the system producing low-oxygen packs but not so much as to cause the machine to have to be slowed down because of product-in-seal problems.

3.13 Packaging materials for VFFS systems

There are many packaging materials that are suitable for use on VFFS systems, from simple polyethylene films for potatoes to complex multilayer laminated materials for products with higher packaging needs. Details of these films are contained in Chapter 7, but here we need to think about the specific issues that films can cause on a VFFS system.

3.13.1 Sealing surfaces

A standard VFFS bag is constructed with the thermoplastic sealing layer on the inside of the bag. This means that when the top and bottom seals are being made, two layers of the thermoplastic are brought together to create a seal. The same goes for the back seal – on a standard bag the two inner surfaces of the material are brought together to form a fin, which is then folded to one side and sealed (Fig. 3.26).

On a VFFS bag with an overlap type of back seal a different situation occurs. The top and bottom seals are the same as the standard bag, with the inner surfaces being brought together to make the seal, but the back seal is different. Here the inside surface of the bag is sealed to the outside surface and this means that the inner and outer surfaces of the bag need to be made of the same material (Fig. 3.27). The different construction of the back seal is dictated by the folding system at the top of the filling tube, so it is important that the correct folding system is in place for the material being sealed. The wrong folding system could attempt to make an overlap seal for a material where the inside and outside surfaces are not compatible and as a result poor seals or even no seals will be made. This is a particular issue on the introduction of new packaging materials into a business where the full implications of moving from a fin seal to an overlap seal have not been thought through and there has been insufficient training of VFFS machine operators. Many businesses will look at the transition from fin seal to overlap seal as a

Fig. 3.26 Here is an example of a fin seal where the inner surfaces of the packaging material are brought together to make a back seal for the bag. Notice also the area where the back seal meets the end seal on the pack. The sealing jaws have to seal through four layers of packaging material in this area and only two across the rest of the end seal. This can give rise to poor seal integrity at the junction of the end seal and the back seal.

Fig. 3.27 An example of an overlap seal where the inner surface is sealed to the outer surface of the packaging material. This can help with a reduction in packaging weight, but the inner and outer surfaces need to be compatible for a reliable seal to be formed this way. Also notice in this picture the area where the back seal meets the end seal of the pack. The sealing jaws have to seal through three layers of packaging material as opposed to four layers with a fin seal.

way of reducing the weight of packaging being used, but this transition must be well managed to avoid costly mistakes.

3.13.2 Printed materials

Many VFFS packages are made with printed materials, and the systems used by VFFS machines must be taken into consideration in making decisions around this. Print quality is nearly always important if products are being packaged for retail sale. Traditional thermoplastics like polyethylene have low sealing temperatures so on the face of it are a good choice for the inner surfaces of multilayer laminated materials for VFFS systems. If, however, the package is to have an overlap seal then consideration needs to be given to the print quality that can be obtained with polyethylene as the outer surface of the pack. Another thermoplastic with better printing characteristics may be the better choice, even if the sealing temperature has to be higher. Finally, in the area of sealing printed films it should be noted that the print inks used will have very different thermal properties from the materials they are printed on. This may cause no problems in normal use, but at very high temperatures when being sealed the inks can change colour and also break down and start to deposit onto the sealing jaws. For this reason, when sealing printed films on a VFFS machine it is vital to keep a close eye on the build-up of materials on the sealing jaws to prevent issues arising.

CHAPTER 4

The design and operation of horizontal form fill seal systems

4.1 Introduction

A horizontal form fill seal (HFFS) system (sometime called a thermoformer) is commonly used in the packaging of solid items that do not flow and so cannot be packaged easily on the simpler and faster VFFS systems (Fig. 4.1). Ham, bacon and sliced cheeses are commonly packaged in this way to produce a pack that is pleasing to the consumer. Hospital instruments are also commonly packed on HFFS systems.

HFFS systems are flexible in that the shape of the bottom part of the pack can easily be changed so that packs of different sizes can be made on the same machine. The bottom part of the pack is vacuum formed on the machine from a flat sheet of thermoplastic material to produce a web of packages that are joined together. After forming, the product is placed into the bottom part of the pack and then the pack is sealed. The top of the pack is made from a flat sheet of material and the top and bottom sheets are sealed together using heat. Finally, the individual packs are cut apart to form the packages for distribution (Fig. 4.2). An HFFS system operates on an intermittent motion basis rather than a continuous motion to allow the different parts of the sequence to be carried out.

4.2 The sequencing of an HFFS system

4.2.1 Thermoplastic material introduced and heated

A reel of flat thermoplastic material is fed into the machine and is pulled through it using a gripper chain (Fig. 4.3). The chain mechanically grips the edges of the film material and moves it forward in an indexing intermittent (stop and go) motion. The base film arrives at a heating station where it is heated using radiant heaters so that the thermoplastic softens and becomes mouldable.

Handbook of Seal Integrity in the Food Industry, First Edition. Michael Dudbridge.

Fig. 4.1 An example of a typical HFFS machine. There are many suppliers of this kind of packaging machine and they are typically used for packaging sliced meats and other foods as well as medical devices. (Picture by kind permission of Multivac UK Ltd.)

Fig. 4.2 A typical collection of HFFS packs. They range from packs with a semi-rigid base made from materials that are around 1 mm thick to packs with quite flexible bases that can be formed from a thermoplastic feed stock that is around 0.25 mm thick as it enters the machine. (Picture by kind permission of Multivac UK Ltd.)

4.2.2 Moulding of softened film

The gripper chains index forward and the softened film arrives at the moulding station. The top and bottom tools clamp together trapping the softened film in between. There is a variety of different ways of forming the bottom of the pack at this stage. Vacuum can be used to suck the softened material down into the base

Fig. 4.3 The base film feed system. The films here are up to 1.5 mm thick so the reels of film are up to 40 kg in weight. Even with the high reel weight, the length of film on a reel can be quite short and so reel changes are frequent compared to processes using reels of thinner films. The base film feed is designed to be low on the machine to minimise the loading issues with such a heavy weight. (Picture by kind permission of Multivac UK Ltd.)

mould. Alternatively, a plug, which is the mirror image of the base mould, can push the softened material down into the mould. The third option is a combination of vacuum and plug assist to form the pack. The third option is usually used where stronger packs are needed and so thicker films are being used.

4.2.3 Cooling and filling

Once the bottom of the pack is formed, the moulding tools open, the thermoplastic is cooled and the gripper chains carry it forward to the filling section of the machine (Fig. 4.4). Here the packs are presented ready to be filled. The filling operation can be manual or carried out by a pick and place robot.

4.2.4 Sealing

Next the gripper chains carry the web of filled packs forward where they are covered by the top film and the packs then move under the sealing station of the machine. Here a heated sealing tool is lowered onto the packs, which are supported by a base tool rising up from under them. The top film is sealed to the base web to form the packs, but the packs are still joined together as they have been formed from a continuous roll of thermoplastic.

Fig. 4.4 The gripper chain is a vital component of an HFFS machine. The base film has to be kept under control even after it has been heated and is quite rubbery. The way this is achieved is by making the film move by gripping it and then applying the movement forces to the chains rather than by pulling the film. Here we can see some packages that have just been formed – notice the gripper chain that runs along both sides of the packages to move them in a controlled way when required. (Picture by kind permission of Multivac UK Ltd.)

4.2.5 Cutting

Next comes the task of separating the packs. This is usually a three-stage process. The first stage occurs when the gripper chains index forward again. Some shaped stamps puncture the packs at the corners where they meet. This creates star-shaped pieces of the bottom and top films. The next index forward takes the packs through a guillotine which cuts in a straight line across the machine. On the final index forward rotary cutters are used to cut the packs apart from each other in the longitudinal direction down the machine. The rotary cutters also separate the packs from the gripper chain so now we have individual packs exiting the system.

4.3 Types of HFFS machines

There are many different types of HFFS systems that all operate the same basic system to manufacture sealed packs, but there are some important differences depending on the required final pack specification and the required packing speed.

4.3.1 Machines with two heating stations

The limiting step on HFFS machines in terms of the speed of the output is usually the heating and softening of the base film prior to moulding. So on machines that are needed to run fast or for packs that need to be made of thick gauges of thermoplastics with a high temperature requirement before they soften, it is possible to incorporate two heating stations into the machine. This will double the potential speed of the machine without having to compromise on the thickness of the bottom film (Fig. 4.5).

Fig. 4.5 The heater section where the film is taken above its Tg (glass temperature) and is converted from a semi-rigid film into one with a rubbery texture. The heaters here have been removed from the machine so that they can be easily photographed. The heating step control is vital to the quality of the final packs and so this is often the rate-limiting step on the machine. If the base film is heated too quickly its texture may be uneven and so faults will occur in the moulding step. (Picture by kind permission of Multivac UK Ltd.)

4.3.2 Machines with two moulding stations and two sealing stations

Some machines are designed with two moulding stations and two sealing stations so that changing from one size of pack to another can be done automatically with no downtime. This can be very useful in a factory where changeovers occur frequently.

4.4 Components and subsystems of HFFS machines

4.4.1 The bottom film handling systems

The bottom films on an HFFS machine can be up to 1 mm thick in order to get the required strength and rigidity into the side walls and base of the package once the thermoforming has occurred. Because the bottom film is thick there is a relatively short length of the film on the in-feed reel, which has the effect of requiring the reel of film to be changed quite frequently. For this reason the reels are as big as possible to prevent the machine running out of film too quickly. The size and inertia of the reels of film mean that the system for moving the film through the machine must be powerful and robust to prevent issues with misfeeding. Once the film has been fed onto the machine, it is gripped by a gripper chain system that controls the film all the way through the machine (Fig. 4.6). The chain

Fig. 4.6 The gripper chains are made up of links, each of which is a spring-loaded gripper. As the chain turns the bend at the start of the machine, small cams operate on the grippers and open them. The base film is guided into position and then, as the chain moves forward, the cam no longer acts and the springs of each gripper force them shut, trapping the two edges of the base film as they close. Because the film is firmly gripped by the chains, the film is totally under control even if it is heated or changes shape. The chain here is moving from right to left and you will notice that one of the links on the chain has been opened ready to have the edge of the base film guided into position before closing again to grip it. (Picture by kind permission of Multivac UK Ltd.)

is made with spring-loaded grippers that control the intermittent movement of the base film. A chain is used to carry out this task so that even when the film is softened in the heater section of the machine the film is perfectly under control and is not subjected to any stretching or pulling forces. The two gripper chains take all the strain of the chain drive system and simply carry the film with them as they are accelerated, moved and then decelerated to a stop at each indexing step of the machine. The gripper chains open to accept the film at the start of the machine and then again at the out-feed end of the machine to release the film. At all other times the base film is gripped and held in a precise and known location so that all of the required operations can be carried out accurately.

4.4.2 The moulding system

This is the part of the HFFS machine where the thermoplastic bottom film is shaped into the required configuration and is able to accept a product. Some products require quite a loose and flexible shape, so thin bottom films are used. Others require a definite shape to help protect or display the product. The moulding section is made up of two parts. First is the heating section. This is where the thermoplastic bottom film is heated to make it mobile and mouldable. Heat is usually delivered from both above and below the film by plate heaters that radiate their energy to the film to increase its temperature. Once the heating time has elapsed, the gripper chains index forward to take the softened film out of the heating section and at the same time deliver more film to be heated. Immediately the softened film comes to a halt at the moulding station the top and bottom tools close around the bottom film and the moulding process starts (Fig. 4.7).

It is worth considering here what is happening inside the mould as the shape of the pack is being formed. The shape of the pack is determined by the shape of the bottom tool in the moulding section. Typically bottom moulds have angled sides and curved corners to aid the moulding process. If the corners of the pack were sharp and square the thermoplastic would be badly stretched resulting in very thin walls in parts of the pack and this in turn would make the pack weak (Fig. 4.8). The heating and moulding operation needs careful control if the bottom film is to be optimally thermoformed. Over- or underheating, time delays or timing issues will mean that the packages made by the HFFS system will not be as required and may well cause problems in distribution.

Flexibility is often designed into the moulding system to allow the HFFS system to produce packs of different capacities with the minimum of downtime due to changeovers. Often the bottom tool is made flexible by the use of inserts to vary the depth of the packs being made without changing the top sealing area size or shape (Fig. 4.9). This means that the sealing tools at the other end of the line do not need to be changed and as a result the machine is back into production very quickly.

Fig. 4.7 The moulding section is at the core of the HFFS machine. Because the moulds are easily changed, an HFFS machine is very flexible and able to produce different shapes, sizes and depths of pack from a flat reel of base film. (Picture by kind permission of Multivac UK Ltd.)

Fig. 4.8 Here is a typical base mould. Notice the angled walls of the mould and the rounded corners to prevent the base film being overstretched during the thermoforming process. Without these features the base film could be made very thin when pulled into the corners and as a result the base of the pack would be weak. (Picture by kind permission of Multivac UK Ltd.)

Fig. 4.9 The flexibility in pack size is often achieved with the use of inserts into the moulds to vary the depth of the packs being made. The top profile shape of the pack remains the same and it is just the depth that is the difference between a 200 g pack of bacon and a 300 g pack. (Picture by kind permission of Multivac UK Ltd.)

4.4.3 The filling station

This is the section of the machine where the product is loaded into the thermoformed bottom film on the machine (Fig. 4.10). This section can be constructed in various lengths and so has a big impact on the overall length of the machine and, importantly, the length of the gripper chain. The chain is responsible for the precise positioning of the bottom film on which the correct operation of the machine depends. We will see in the maintenance section how important the length of the chain is to the overall performance of the machine in the long term. If the chain is stretched then the relative position of the moulding section and the lidding and cutting section – for example the guillotine position – will be compromised with the consequence that the packs will be in the wrong position when cut. Precautions need to be taken in this area to ensure that the seal areas of the base film do not become contaminated by the product being placed in the packs. Drips of liquid

Fig. 4.10 The filling section is where the product being packed is placed into the newly formed base of the package. This is also the section where seal integrity issues can arise through seal area contamination. Product that is not positioned correctly will contaminate the seal area and when it is time to seal the pack with a top film the seal will fail. For this reason many companies are automating this section of the machine with the use of precisely controlled delta robots rather than using people to carry out this work. (Picture by kind permission of Multivac UK Ltd.)

and smears of product on the seal area would be sufficient to cause seal integrity problems for the final packs, so this should be prevented if good packs are to be manufactured. The other issue that can cause problems when it is time to seal the packs is if the product is deeper than the moulded shape it is sitting in. If the product is 'proud' of the top surface of the packs there is no facility on an HFFS machine to correct that and the lidding section will not be able to place a flat sheet of the top film over the packs. The film will, inevitably, become creased and this will cause the seal integrity of the packs to be poor, with leaks occurring when any crease in the top film crosses the seal area.

4.4.4 The lidding section

This is where the top film is fed onto the machine and is sealed to the base film by the lidding tool (Fig. 4.11). The top film will also contain printed information, date codes, batch numbers and the like, so this section can be quite complex for what appears, at first, to be a simple task of unwinding a reel of film and feeding it onto the machine. Once the top film has been placed over the filled packages and the positioning of any printed information is correct, the packs, on the next indexation of the gripper chains, will pass into the sealing section. Once the packs

Fig. 4.11 A typical lidding section of an HFFS machine where the top film is brought into position above the filled packs by a film handling system that is designed to precisely control the film so that it does what is expected. Handling any thin film in a factory environment can be tricky and the film can start to wonder off line or not move in the predicted way unless it is carefully controlled. The unwinding of the top film from the reel and the precise threading of the film through the machine are vital if the correct control is to be maintained. With a mistake in this area of the HFFS machine it is possible that the surface of the top film that was supposed to be on the outside of the finished pack could end up being the surface that the machine is trying to seal to the base. (Picture by kind permission of Multivac UK Ltd.)

have stopped moving, the top and bottom tools will close around the packs and a heat seal will be made. The sealing temperature and dwell time will be controlled by the machine operator and should be within the tolerances recommended by the packing material supplier. This will ensure a consistent seal and prevent too high a temperature being used and damage being caused to the packaging materials. There is a need to ensure that the seal is occurring in exactly the correct position on the pack and often there is a machine adjustment that can be made to optimise the position of the sealing section relative to the position of the moulding section at the start of the machine. During the sealing operation it is quite common to engineer in an easy-open feature. This is usually achieved by designing the shape of the sealing tool to create a tab in one corner of the pack.

4.4.5 Labelling of the packs

An ideal time to apply adhesive labels to the packs is once they are sealed but before they are cut. The packs are all in a precise and known position, so it is relatively easy to automatically apply labels to both the top and the bottom of the packs if required (Fig. 4.12).

Fig. 4.12 The application of labels to packs is a precise function of many HFFS machines. Without careful maintenance and operation set-up the label applicators can place large forces onto a pack as the label is applied. This is occurring very soon after the top seals are made on the pack and the thermoplastics will not have fully cooled back to their stable temperatures. It is possible for a wrongly operated label applicator to damage a perfectly good seal made a few seconds before. (Picture by kind permission of Multivac UK Ltd.)

4.4.6 The cutting section

The packs arrive at this section of the machine as a continuous belt linked together and they need to be separated into individual packs. This needs to be done precisely to ensure that the packs look neat and that the position of the cuts does not have an impact on the integrity of the seals. The first step in separating them is to stamp out small star shapes from the intersections of the packs. These shapes will form the curved corners of the packs. This needs to be done where the base film used to mould the bottom of the pack is quite thick and rigid. The top of the pack will not have been stretched during the moulding process and so will retain the full rigidity of the original base film. This can be a thermoplastic of up to 1 mm thick and so, once cut, could cause damage to people handling the packs or to carrier bags in a retail outlet. There are few things more frustrating at the end of a long shopping trip than the carrier bag you are using splitting open because of a sharp corner on a pack of ham or pre-sliced cheese. Once the corners have been formed, the packs are indexed forward by the gripper chains and a guillotine is used to cut between the packs across the width of the machine (Fig. 4.13a). This separates the packs into strips but they are still under the control of the gripper chain to which they are still attached. The final part of the cutting section, and the

(a)

(b)

Fig. 4.13 The guillotine systems **(a)** and rotary cutters **(b)** on HFFS machines need to be set up precisely in relation to the position of the moulding section of the machine. A misaligned rotary cutter can, instead of simply separating the packs, cut into the seal area of a pack and so cause seal integrity issues. The bigger problems in this section of the machine come from the guillotine cutter. It needs to be set up to match the indexing of the machine, but occasionally the ideal position varies because of gripper chain stretch issues on the machine. The guillotine position ends up being a compromise where its actual position varies on either side of the optimum position. If the chain stretch issue becomes too bad then the guillotine can cut into the seal areas on the packs and so cause integrity problems. (Picture by kind permission of Multivac UK Ltd.)

final part of the HFFS machine, is the longitudinal cut. This is usually achieved using rotary cutters that cut the thermoplastic between the packs and also cut the edge packs away from the gripper chain (Fig. 4.13b). The packs are now separated and they are usually moved away from the machine using a conveyor belt or sometimes they drop down a 'ski slope' and onto the next part of the packing process.

4.4.7 The vacuum system

This is a vital component of an HFFS machine. It is the vacuum system that is used to draw the softened thermoplastic film down into the bottom mould. This is sometimes assisted with a 'plug assist' system to push the film down from above, but it is the vacuum that ensures that the desired shape of the bottom of the pack is achieved. In most HFFS machines the vacuum pump is 'on-board' the machine, but in some cases the vacuum source pump is located away from the shop floor and is connected to the HFFS machine with pipework. The advantage of an on-board vacuum pump is that it is close to where the vacuum is needed and there will be only be a small pressure drop in the pipework. A remote vacuum pump may help keep noise levels down but there will be a pressure drop in the pipework that will make this option less efficient than the on-board option. The vacuum pump assists with the moulding operation by drawing the rubbery thermoplastic base film down into the mould. Small holes are carefully positioned in the base of the mould to get the base film to flow into the correct positions.

4.5 Setting up an HFFS machine

There are general set-up procedures that ensure that the HFFS machine produces good quality packs of the correct shape. There are some critical areas of the machine that if not set up correctly will lead to waste and poor quality and maybe even machine damage.

4.5.1 Temperature control of the heaters in the moulding section

The heaters in the moulding section have the task of softening the thermoplastic of the bottom film sufficiently for it to be formed into the desired shape. The heating section of the HFFS machine is designed to heat the film from above using radiant heaters. The gripper chains bring the film into the heating section and then stop for an exact amount of time. The radiant heat starts to soften the film as energy is transferred from the heater plates. Radiant heat transfer is subject to a rule of physics called the inverse square rule. The distance from the surface of the heating element to the thermoplastic film needs to be as small as possible to optimise the heat transfer. The quantity of heat transferred is proportional to the inverse square of the distance. This means that if the distance doubled, the heat transferred would be divided by four, if the distance tripled, the heat transferred would be divided by nine, and if the distance quadrupled, the heat transferred would be divided by 16. So the control of distance between the heaters and the thermoplastic is vital and this can be difficult when the item being heated is a thermoplastic that can distort and bend under the influence of the heat you are trying to transfer to it.

If the temperature of the film is not high enough when it enters the moulding section the film will not flow correctly down into the corners of the mould. The film will not flow correctly even though the vacuum is still pulling it down into the mould. This can result in the thermoplastic material being very thin in the corners of the pack, with the plastic material that should be in the corners remaining in the side walls of the pack. If the temperature of the film is too hot then the film will move very easily to the base of the mould and as a result the side walls will be thin and weak. In the worst case, the walls or the corners can be so thin that they are easily damaged and leak. It is often the case that on start-up it can take a few seconds for the temperature of the heaters and the film to become stabilised and for the packs to be perfectly formed. The key to a good set-up is to ensure that

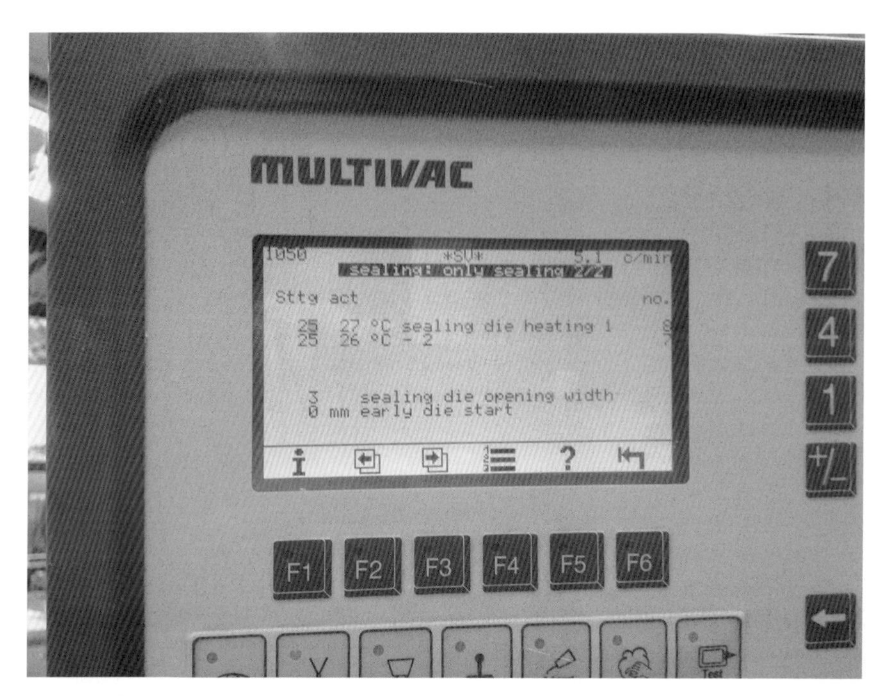

Fig. 4.14 The use of precise temperature control is required if consistent quality of moulded packs is to be produced. Different techniques have been developed over the years to improve pack consistency. Among the favourites is a technique called 'plug assist' where the base film is physically pushed using a male pattern of the mould to ensure the thermoplastic moves in the correct way. (Picture by kind permission of Multivac UK Ltd.)

the minimum amount of waste possible is made during this phase of the start-up of the machine. As the base mould and plug assist start to work they will increase in temperature by absorbing heat from the bottom film. Once equilibrium is reached, the mould will be gaining and losing heat in the same quantity on each index of the machine, so the behaviour of the film and the packages will become more consistent. The set-up and control of thermoforming temperatures are key to the success of the packing operation and these need to be included in standard operating procedures for the machine to ensure that faulty packs are not manufactured (Fig. 4.14).

4.5.2 Positioning of the sealing heads

The next set-up task on the HFFS machine is to check the position of the sealing heads relative to the moulding station. As the packaging indexes forward on the line it is important to ensure that the sealing head is positioned so that it seals the packs in exactly the correct place. This is usually achieved by the use of a set of gauges mounted on the machine (Fig. 4.15). These indicate to the operator if the position of the sealing head needs to be adjusted to be in the

Fig. 4.15 Gauges are used to help set up an HFFS machine and get all of the sections of the machine into the correct locations relative to each other. One of the main issues with HFFS machines is the tendency for the gripper chains to stretch unevenly, as a result of which the positions are sometimes a compromise. (Picture by kind permission of Multivac UK Ltd.)

correct position. During commissioning of the machine, readings will be taken from the gauges for each size and shape of mould so that when that mould is placed onto the machine again it is relatively easy to check and adjust the position of the sealing head. If the head is in the wrong position it is possible that the seals will be so far out of position that they will be weak or will even fail completely.

4.5.3 Positioning of the star stamps, guillotine and rotary cutting knives

The third big set-up task on a thermoforming HFFS machine is the positioning of the star stamps, guillotine and rotary cutting knives relative to the position of the packs as they appear at that end of the machine. As with the set-up of the sealing heads, all of these positions are normally determined by gauges and positional markings on the machine. Settings are recorded when the machine is commissioned and these are used to configure the machine correctly for each size and shape of pack. Misalignments here can cause weak and faulty seals, especially as the seal area is usually only 3 or 4 mm wide, so even small misalignments can have big impacts on seal quality.

4.6 Maintenance

The hygiene and maintenance of an HFFS machine is critical to its performance. There are numerous parts of the machine that have an impact on the seals of the pack. If faults in any of these go uncorrected they will fail to deliver what is required and the machine will produce packs that may leak.

4.6.1 The gripper chains

The first and most important area to consider is the gripper chains (Fig. 4.16). These chains run the entire length of the machine and back, so a machine of, say, 10 m would have two chains of around 21 m length each. A typical gripper chain has a link every 1–2 cm, so that's around 2100 links in each of those two chains. When an HFFS machine is in use the machine motion is intermittent, so each time the machine starts to move the drive system has to overcome the inertia and

Fig. 4.16 Wear on the gripper chains can cause uneven stretch and as a result the positioning of the packs becomes less precise. When a typical seal is 3 or 4 mm wide it does not take much for this seal to be made in the wrong position and this can cause a problem for seal integrity. (Picture by kind permission of Multivac UK Ltd.)

frictional forces in the system and then stop everything again within 20 cm or so. This action subjects the gripper chains to large forces and this has a tendency to stretch the chains. The chains most commonly stretch because the pins that join the links of the chain together wear slightly. The links themselves can also elongate marginally. You will notice that a small amount of wear at each of the 2100 links in the chain can soon result in the chain elongating. The problem of the chains stretching on an HFFS machine can be made worse by poor hygiene practices. On its return leg near the bottom of the machine a section of the chain can be left wet in sanitiser overnight (or until the next shift starts). This can have the worst possible consequences for an HFFS machine. The stretch of the chain can be made uneven by the action of the cleaning chemicals while the machine is stopped. Part of the chain can be more stretched than the rest and this can make the correct set-up of the machine almost impossible to achieve. The relative positioning of the sealing heads and cutting guillotine are based on knowing the exact length of chain that is indexed with each stroke of the machine. If the same number of links can measure different lengths then the position of the sealing heads and guillotine can, at best, be a compromise. The sealing head and guillotine would have to be set up to be slightly in front of the correct position in a compromise to accommodate both the stretched and the unstretched sections of the chain. The sealing area of a pack produced on an HFFS machine is around 4 mm wide, so the tolerances for the chain are very high to ensure that each pack stops in the correct location to be sealed and cut. With 2100 links per chain it would only take a stretch of around one thousandth of a millimetre per link to render the position of the sealing and cutting 50% out. Two thousandths of a millimetre at each link would mean that the seal and cut were being made in totally the wrong position. This misalignment can lead to seal problems and be the cause of a large amount of waste product.

4.6.2 The sealing heads

Maintenance of the sealing heads is also important to ensure that the desired seal is created. If the sealing heads are dirty or are covered with carbonised materials then there could be areas within the seal that are weak or that are not made at all (Fig. 4.17).

4.6.3 Seal area contamination

The biggest source of faulty seals on HFFS machines is the seal area contamination that occurs as the product is being loaded into the pockets made in the bottom film. Maintenance around ancillary machines is vital to keep the seal areas clean prior to the top film being positioned ready for sealing. Dripping depositors or misaligned placement systems can have a very poor effect on the seals the machine is making. Maintenance and adjustment of these systems will ensure a large increase in seal area cleanliness and as a result a higher proportion of packs will be sealed correctly. Occasionally, the task of loading products into the HFFS is fully automated and on these occasions the issues already discussed about the impact of uneven gripper chain stretch become important. Packages are being

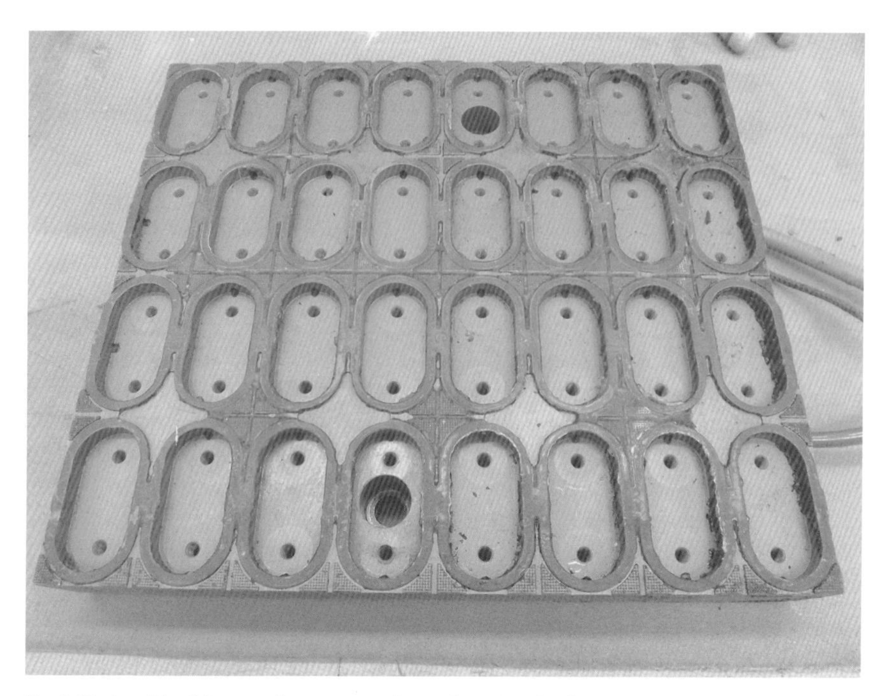

Fig. 4.17 As with all heat sealing systems the surfaces involved run at high temperatures. If the sealing surface becomes dirty then the contaminant will burn and carbonise on the surface. Here we see a heating surface that has surface carbon and as a result the seals being made here are likely to show seal integrity issues. (Picture by kind permission of Multivac UK Ltd.)

designed to be as small as possible to reduce the weight of packaging materials used, so tighter and tighter tolerances are being used. The precision made possible by a robot loading system or other type of automation is far superior to that of a human operator loading the machine. Humans loading an HFFS system will cause seal area contamination and this, in turn, will cause leaking packs. If human filling operations are to be used then the operatives should be given regular job rotation to try to minimise their fatigue and optimise their accuracy and the care with which they load the machine (Fig. 4.18).

4.7 Special case of modified atmosphere packing

HFFS machines are often used for modified atmosphere packing (MAP) (Fig. 4.19). This is where the air inside the package with the product is taken away and replaced with a modified atmosphere. This is done to restrict the growth of micro-organisms or reduce the deterioration of the food by restricting the quantity of oxygen in the pack. Occasionally, high levels of oxygen are used (40%) to enhance the appearance of red meat products such as beef burgers or beef steaks.

Fig. 4.18 The contrast in performance between a human loading an HFFS machine with product and a delta robot doing the same task is marked. The delta robot will work to the same level of precision all of the time. A human operator may not be so precise and will certainly be more variable as fatigue builds up during a shift. (Picture by kind permission of Multivac UK Ltd.)

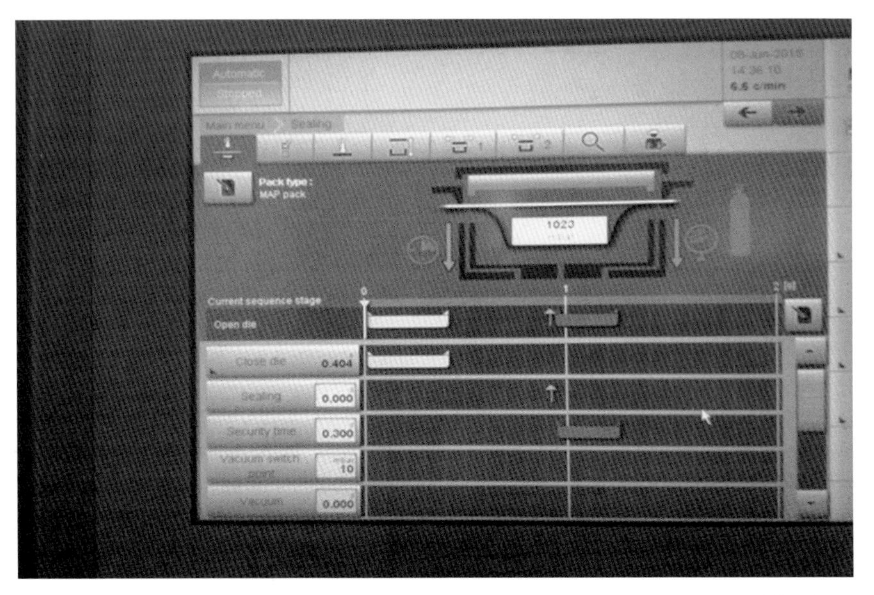

Fig. 4.19 A modified atmosphere HFFS machine is very common in the packaging of food products where the MAP process improves the shelf life of the product. MAP is very reliant on good seal integrity, so when it is used the HFFS machine should be maintained and operated at the highest levels. (Picture by kind permission of Multivac UK Ltd.)

The modification of the atmosphere on an HFFS machine is carried out on the sealing section of the machine. At the start of the line, when the gripper chain is fed with the bottom film, additional slots are cut into the bottom film. The slots form the path by which the modified atmosphere enters the packs just before sealing. As the top film is lowered onto the full bottom film below the modified gas ports are opened and the gas flows through the slots in the bottom film and into the packs. This flow of gas is used to push out the normal air in the pack. Once the gas has flowed for the desired amount of time, the sealing sequence is initialised and the modified gas is sealed into the pack. Some HFFS machines operate a method of MAP called Vac/Gas. This is where the top and bottom tools come together in the sealing section and form a sealed chamber. A vacuum pump is started to draw the air out from the chamber (and the packs inside the chamber). Once the vacuum cycle is complete, the modified atmosphere is released into the chamber. Once the modified atmosphere is inside the packs, the sealing sequence is initiated and the modified atmosphere is sealed into the packs.

The key factor here is that when a modified atmosphere is used inside the pack to help protect the contents then a very good seal is required to keep that atmosphere inside the pack and to stop it leaking away.

4.8 Films used

The thermoplastics used in HFFS machines have a variety of tasks based on the requirements of the final package. If the pack has to be quite rigid then thicker bottom films will have to be used. Thicker bottom films take more energy to soften, so it is likely that a material such as polypropylene will be used because it softens at a lower temperature than some other thermoplastics. There are other options, including thinner base films to produce a soft pack rather than a rigid one. It is common for medical instruments to be hygienically sealed inside a flexible pack where the top film is a paper-based laminate. So the seal is made between two thermoplastics but one of them is bonded to a paper layer to give the pack some rigidity and also to make it easier to print instructions onto the pack.

Much more consideration will be given to thermoplastic materials in Chapter 7.

CHAPTER 5
The causes of seal integrity issues

5.1 Introduction

This chapter will review a study carried out for the Waste and Recycling Action Programme (WRAP) in the United Kingdom. The study looked at seal integrity performance on over 100 sealing systems and the statistics produced identified the main causes of seal integrity issues. The study focused on packaging systems within the food industry but the lessons learned apply across all heat sealed flexible and semi-flexible packaging systems.

5.2 Background

Following a proposal to reduce food waste through improved methods of monitoring and managing seal integrity on food packaging, a team from the University of Lincoln's National Centre for Food Manufacturing was asked to quantify the size of the food and packaging waste issue associated with faulty food packaging seals during the packaging process.

Due to the sensitive nature of the topic there is much anecdotal evidence of the problem but very little published data. However, preliminary investigation revealed that some companies had some records used for internal performance measurement, although these were often incomplete.

The University of Lincoln was able to gain access to a number of food processing factories where interviews and measurement provided some insight into the issue. Access was granted on the strict understanding that the companies would remain anonymous.

Handbook of Seal Integrity in the Food Industry, First Edition. Michael Dudbridge.

5.3 Objective

The objective of the work was to quantify the amount of food and packaging waste associated with faulty seals created in the food packaging process. In addition, the project aimed to identify the amount of potential food and packaging waste created by faulty seal integrity that can be reworked on site, how much has to be scrapped and how much continues along the food chain to food retailers and householders.

A further set of objectives related to describing how seal integrity is monitored and to identifying activities associated with best practice in the management of the food packaging process.

5.4 Study methods

The study set out to produce quantitative data to meet the objectives. The study team were given very good access to managers at various levels within the factories and as a result were able to collect qualitative data including opinions, plans, relationships to environmental issues and so on. Some of this data is recorded and some of it informs the work of the team.

The study consisted of two elements: an interview with senior managers at the factory site and a visit to the factory shop floor to observe and measure aspects of seal integrity.

A composite data collection document was developed in conjunction with WRAP. The data collection document was modified following an initial pilot study in one factory to improve its ease of use and the quantity of data collected. This was supported by test measurement documents to record vacuum testing and dye penetration testing of packs removed from the production lines.

The key data items that were sought related to the companies and the individual packaging lines. The measurements were made on samples taken from each production line using a standard vacuum testing rig and a dye penetration analysis system.

5.4.1 The interview

A senior factory manager was interviewed and background data to the project was collected. This included factory turnover, number of employees and number of packs produced per day.

This interview obtained the company's commitment to the work and identified key factory personnel to assist with the shop floor study. General information about the company was also collected.

5.4.2 Shop floor study

Production lines were visited over a period of several hours for data collection to occur. The factories used in the study were, almost without exception, divided into high-risk and low-risk areas. Data on packaging machines was collected during visits to the high-risk areas and seal integrity testing occurred in the low-risk areas. Waste generation was monitored wherever it occurred in the packing process.

5.4.3 Vacuum testing

Vacuum testing was carried out using a chamber machine. The packs were exposed to increasing levels of vacuum until it became apparent that a fault in the seal had occurred. This gave information regarding the initial seal integrity of the pack and also the seal strength and its ability to survive the rigours of the supply chain.

5.4.4 Dye penetration testing

Dye penetration testing occurred using methylene blue dye mixed with a small quantity of detergent to decrease the surface tension and increase the dye's ability to find small seal anomalies. The dye test results were recorded using high-resolution photographs.

The shop floor study was, inevitably, a snapshot of sealing performance but was supported by other data gained during the work.

5.5 Companies studied

The effectiveness of a company in achieving good sealing on its food packaging is extremely important. The number of faulty packs that might be released into the food supply chain is highly sensitive and can affect the business relationships between suppliers and customers. For these reasons it is necessary to maintain the confidentiality of the participating companies and the data collected.

Initial discussions with the food industry identified that five companies, representing a wide range of packaged food, were willing to assist with the project by making data available and allowing access to factory procedures at 10 sites. In reality we obtained the cooperation of six companies representing 11 factory sites.

It was important to ensure that the factories and companies in the sample represented a fair view of the food industry. Four factors were taken into account.

- Product category – frozen, chilled, ambient
- Product constituent – meat, vegetable, both
- Size of factory – measured by turnover
- Intensity of production – measured by number of packs produced per day

5.5.1 Category of food product

The categories were designated as chilled, frozen and fresh (ambient) products. The diagram in Fig. 5.1 shows the proportion of each in the sample. Most of the packaging lines were packing chilled products, which are the most vulnerable to leakages in terms of food safety and organoleptic properties. This is especially so when the food is packed in a modified atmosphere.

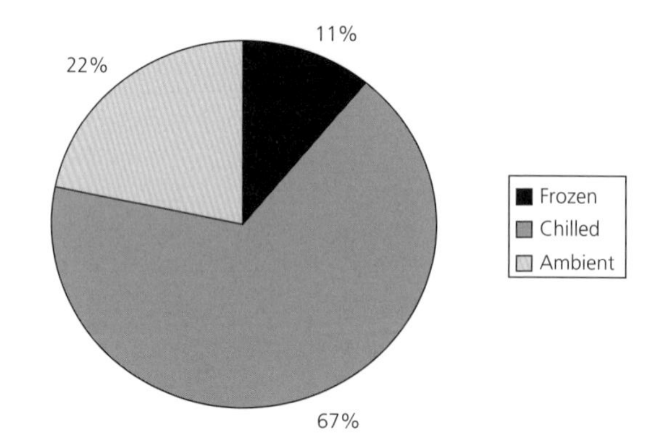

Fig. 5.1 Proportion of product type packing lines.

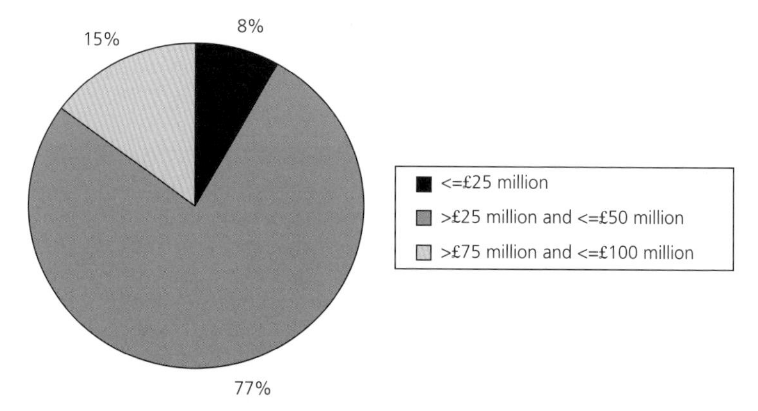

Fig. 5.2 Factory size by turnover.

Fresh produce often requires packaging merely as a container for the product and to protect it from physical damage. Frozen foods share this as a major requirement.

5.5.2 Product constituents

Subdivision of the above categories shows that the products at the factories investigated included meat products, vegetable-based products and products containing both constituents. Factories were also selected to ensure that the test sites represented products containing liquids, sauces, particulates and powders.

Finally, factories were selected that represented all of the major types of heat sealed food packaging types. Vertical form fill seal (bag makers), tray and pot sealing, horizontal form fill seal (pack makers) and flow wrapping systems were all represented in the factories and were surveyed during this study.

5.5.3 Size of factory

There was a range of factory sizes to represent the various scales at which food production is carried out in the United Kingdom (Fig. 5.2).

5.5.4 Intensity of production

This consists of the number of packs produced per week. This measure correlates strongly with turnover but not with number of employees. It is a further indicator of the size of the factories. It is an important output measure which indicates the importance of even small proportions of packages lost through inadequate sealing. Many of the factories we surveyed were producing over half a million packs per week.

5.6 Key results

The major objective of this work was to identify the following.

- The proportion of packs detected as having seal integrity issues
- The estimated quantity of food that had to be disposed of because of seal failure
- The estimated quantity of food that was repacked following seal failure
- The estimated quantity of packaging material that had to be disposed of because of seal failure

In addition to meeting these objectives, we were able to test a number of hypotheses as to why this was happening and to identify good practice in maintaining seal integrity. We were also able to test a sample of packs using both vacuum and dye penetration methods to begin investigating the benefits of continuous versus sampling-based seal integrity monitoring.

5.7 Rate of seal failure

When confronted with the question of the rate of seal failure most companies were able and willing to give an answer. It often took some time to identify this parameter as several data sources often had to be accessed. While the average number of seal integrity problems over a year is usually small (about 1% of production in the sample group), this represents the percentage of packs that the current methods of checking manage to detect.

When the study team asked factory managers about customer complaints relating to seal failure, numbers of around 1.5 to 2 per 100 000 packs were revealed. This indicates that even with considerable effort going into checking for 'leakers' at the factory, some make it into the supply chain. Anecdotal evidence and observations in food retailers suggest that the problem is much higher than the 2 packs per 100 000 suggested by customer complaints data.

Further evidence of the scale of the problem was shown by the dye testing and vacuum testing carried out in the factories as part of this study.

This work concludes that the vacuum test (Fig. 5.3) was less able to detect leaking packs than the dye test (Fig. 5.4). Of all packs in the dye test, 37% showed leaks, while for vacuum down to 750 mBar leaks were detected in only 26%.

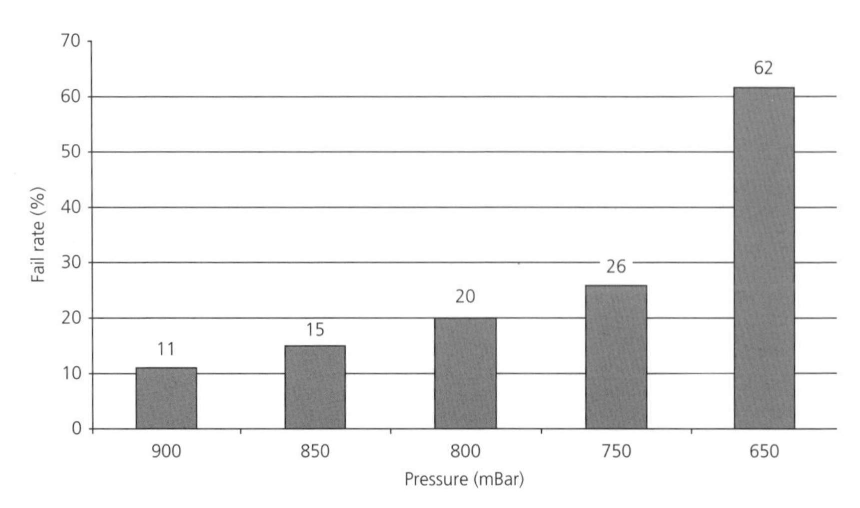

Fig. 5.3 Vacuum tests of food packs. Packs were subjected to vacuums of different levels to test them for their seal integrity.

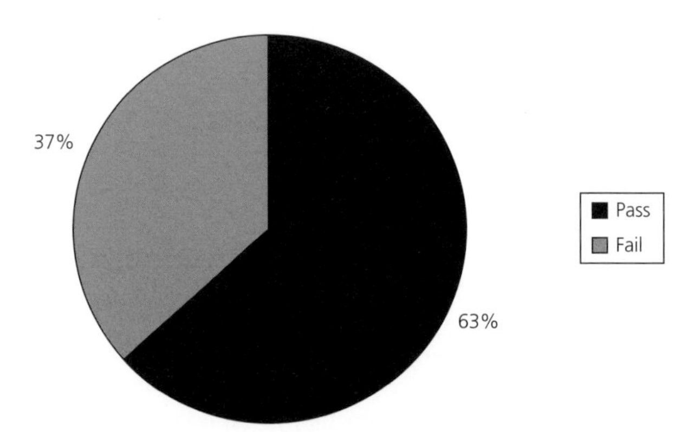

Fig. 5.4 The dye penetration test was used as part of the test procedure for this study. Dye penetration testing did not form part of the normal seal testing procedures at any of the factories in the study. The management at all of the factories were interested in the test procedure and how it might fit into their quality management procedures.

At a very high vacuum of 650 mBar the test was so severe that even some well-sealed packs were destroyed, so the failure rate increased to 62%. It became apparent that different testing methods were being used throughout the industry to judge the adequacy of package seals and that there was little standardisation of the testing equipment or the testing procedures, even on a single factory site.

This further supports the belief that potentially leaking packs are being released into the supply chain. The actual quantity of food and packaging wasted between factory and final consumer will be dependent on the rigours of the distribution channels.

5.8 Discussion of the impact of poor seal integrity

Minimising the problems of seal integrity is of great importance to food companies for a number of reasons.

- The problem is often severe for a short period rather than running at a constant low level. This causes disruption to the whole food packaging process and can have consequences upstream for processing areas of the factory.
- It can often cause a shutdown of the production line.
- Penalties for letting 'leaking packs' reach the customer are severe.
- The cost of monitoring at the factory can be high.
- Using manual methods of monitoring often gives inconsistent results.
- Products that are packed in modified atmospheres rapidly lose shelf life when exposed to even small quantities of air.
- Packed food represents the stage of production where considerable value has been added to the raw material and therefore is a greater cost to the business than food products rejected earlier in the processing chain.

Seal integrity issues are caused by two main routes. Most commonly the seal is faulty because of food entrapment in the seal area preventing a good seal being made. The second route is a fault on the sealing machine causing the seal to not form correctly or to be very weak.

The importance of reducing the apparently small average number of leaking packs with regard to reducing food waste and environmental degradation is in the vast volume of packaged food produced each year.

The total quantity of food consumed in the United Kingdom in a year is about 18 million tonnes. The 1% rate that the factories detect is therefore equivalent to at least 180 000 tonnes of food waste per year. The impact of this large quantity of waste is often minimised by rework in the factories but this too has an impact on manufacturing costs and is by no means possible or practicable in all cases.

This is the quantity of waste identified at the factory and sent to landfill. But the impact of inadequate seals carries on down the supply chain. Of the packs that we tested using the dye penetration 37% test showed a tendency to leak and it can be assumed that a proportion of these will be insecure in the food distribution system or in the home. Without further laboratory testing to simulate the rigours of the supply chain or a survey of the food retailers we think that 10% of the weakened packs might well fail to provide sufficient protection for the food. Therefore a further 666 000 tonnes (18 million tonnes × 37% × 10%) of food and associated packaging could enter landfill after leaving the food factories because of seal weaknesses. Other work by the researchers has indicated that waste in retail outlets because of seal faults could be as high as 10 packs per store per day.

Using these figures, the potential seal integrity problem at the factories is not the 1% of production that is captured but 4.7% of production including the packs that reach the distribution chain.

5.9 Disposal and repacking of food

The survey found that on average 60% of the food from damaged packs was disposed of and 40% was repacked. The split depended very much on the industry sector, with some sectors repacking no food and others able to repack a fairly large proportion. This is a consequence of strict adherence to 'high-risk' and 'low-risk' areas within food factories.

Final packaging usually occurs in a high-risk environment where, in order to reduce health risks, the food will not be subject to further processing. In addition, it is usually in the form in which the customer expects to receive it and any further processing will reduce quality, or at least the perception of quality. Once sealed into the retail pack, the food generally moves out of the high-risk area to a low-risk area where it is packed into crates or outer cases. The move from high risk to low risk is usually rapid and automatic. It is usually in a low-risk area that a seal integrity issue is first detected. It is usually difficult to repackage food at this point in the chain and for this reason it must be disposed of safely. This usually means to landfill.

5.10 Disposal of packaging

This project did not set out to measure the quantity of packaging waste due to poor sealing performance but it was recognised that, once the food was inside a pack and the pack was rejected for poor seals, none of the packaging material could be recovered. However, in all the factories measures were being undertaken to separate recyclable items from those that were not. The most difficult separation procedure was felt to be separating food stuff from packaging in the final product.

5.11 Managing the sealing activity

Managing packaging and sealing is one of the most important activities in food manufacturing.

As part of this study we questioned food company managers about how they managed the procedure. The three major ways in which the process is managed are as follows.

- Methods for monitoring the integrity of seals
- Management of the standard operating procedures
- Maintenance procedures

5.11.1 Methods for monitoring the integrity of seals

Observation of the packaging lines and discussion with team leaders established how the sample companies monitored and maintained the quality of the food packages (Fig. 5.5).

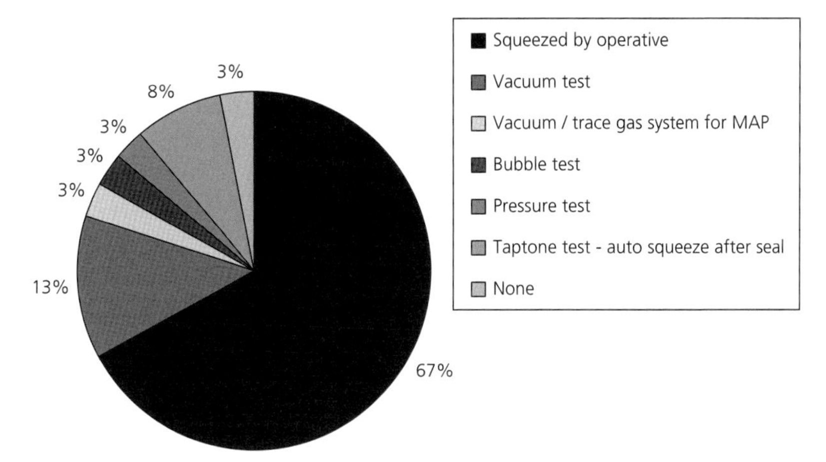

Fig. 5.5 Seal integrity monitoring methods used in the factories studied.

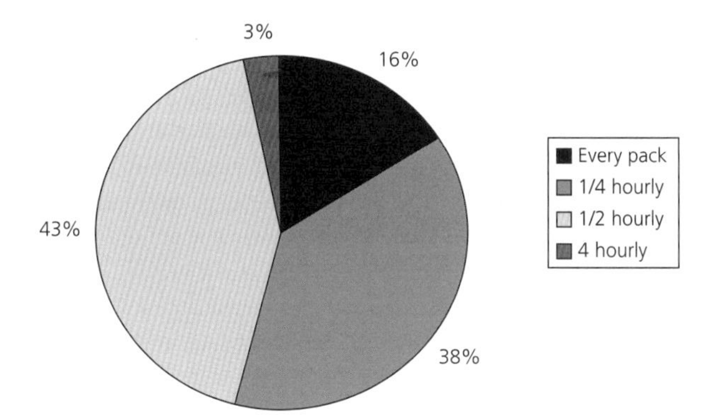

Fig. 5.6 Frequency of seal integrity testing is something that is selected in a factory. There are no standards for this. Because principal cause of seal integrity issues is product entrapment in the seal area, sampling is going to have to be at a very high level to catch what is essentially a random event. Later in the book we look at 100% inspection systems, which are the only real solution to this problem.

The most common method of monitoring the integrity of the seals is by manual squeezing of the packs. If the pack appears to leak it is rejected. Previous investigations available to us indicate that the precision of this method could be questioned, while controlled experiments have shown that low-level leaks are difficult to identify. In Chapter 6 of this book we will look in detail at the available testing methods and equipment for the detection of faulty seals.

Manual methods of seal inspection rely on all production workers being vigilant and on an informal method of deciding when corrective action is needed and what that should be. When a more formal method of monitoring is in place, the most common sampling interval is half hourly, while checking every 15 minutes was done in 38% of the sample and 16% claimed to monitor every pack (Fig. 5.6).

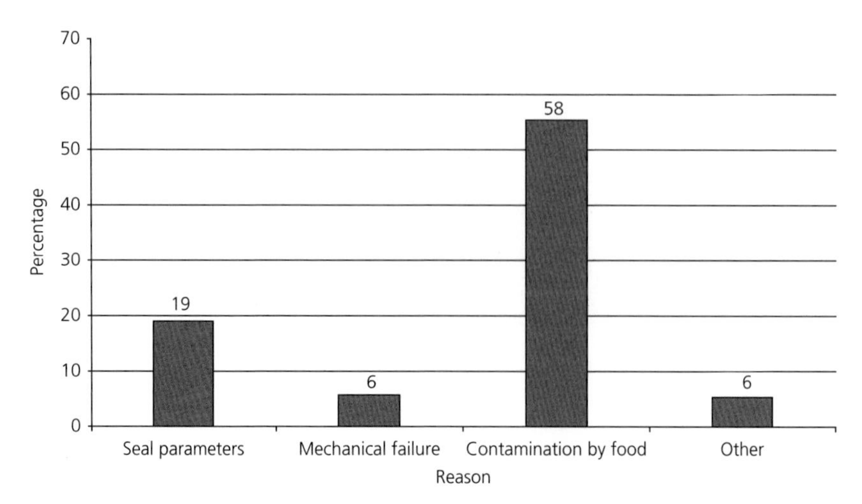

Fig. 5.7 Reasons for seal failure. Here we see the main reasons why seals fail. Working on these issues will improve the performance of your sealing machines considerably.

The intensity of sampling was related to the importance of maintaining product in its packaging, for example products in modified atmospheric packaging. Factories that conducted good seal integrity monitoring did so because their product was reliant on good seals for food safety or quality concerns.

The reason for seal failure was in most cases 'contamination by food'. Problems with setting the parameters of the sealing units also figured, but such problems usually lead to systemic failure which may be more easily identified (a continuous series of failed seals) than the more random nature of food contamination (Fig. 5.7).

It is clearly important to monitor the integrity of seals. While this study cannot establish clear statistical correlations between monitoring methods and reduction in package seal problems, observation and discussion with the managers identify a clearly defined monitoring system as a key element in good practice for food packaging.

An interesting observation during the testing was around the issues of seal peelability. Consumers like the convenience of easy-to-open packs with peelable seals, although the definition of peelable is a matter of opinion and difficult to quantify. No evidence was seen in any of the factories of a 'peelability' test. Seal strength and peelability are closely related and this is a possible area for future work. Chapters 8 and 9 of this book look in detail at this area.

5.11.2 Management of the standard operating procedures

Lean manufacturing theory would suggest that standard operating procedures should lead to more consistent and more effective sealing of food packs. The study team gathered some data on how often the operating procedures for the packaging machines were changed, for example changes in dwell time or sealing temperatures.

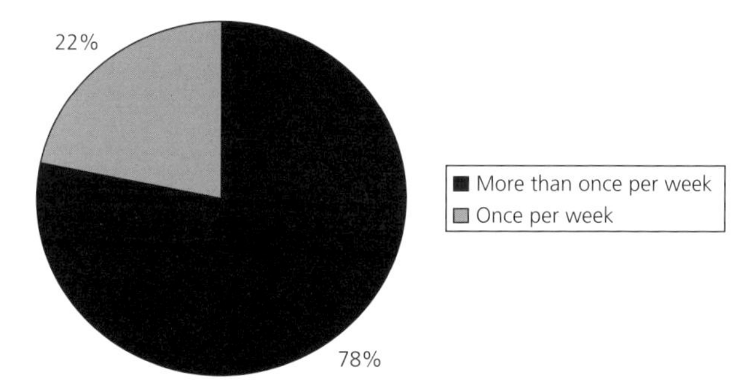

Fig. 5.8 Frequency of changing operating procedures is a good indication of the stability and robust nature of the standard operating procedures for packaging machines.

Eighty per cent of the packaging lines were adjusted more than once per week. The remaining 20% were adjusted at least once per month (Fig. 5.8). This level of micromanagement is related to team leaders making instant decisions often without much data to support those decisions. We feel that if there were a better link between the monitoring function and the adjustment, that is, a better feedback loop system, better seal integrity (and other aspects of packaging management) could be achieved.

The factories with more robust standard operating procedures for their packing systems appear to have fewer seal integrity issues with their packs. A disciplined approach to the operation of sealing machines was the key to better seal integrity performance.

5.11.3 Maintenance procedures

Maintenance procedures are thought to be crucial in achieving efficient and effective sealing of food packs.

To investigate whether good packaging could be at risk from inadequate maintenance procedures we asked about systems employed. About 50% of the packaging lines used planned preventative maintenance (PPM). This is judged to be a low proportion. While there is insufficient data to establish clear, statistical evidence relating poor sealing to lack of PPM, the better performing packing lines seem to use planned maintenance systems.

One important aspect of maintenance is calibration of the sensors in the sealing system. Sensors for temperature can drift in their measurement accuracy and this can have several effects in the packaging process. First, the machine operator will make a machine adjustment to try to improve the seal quality. The 'go to' response to a poor seal is to increase the sealing temperature to try to recover the seal to an acceptable level. The temperature is often increased to a point where the packaging materials are becoming damaged by the high temperatures. A robust calibration checking system will highlight the actual seal temperatures being used and the operator will be able to see that the temperatures are not normal.

A second important aspect of maintenance for good seals is really a hygiene and cleaning issue. To make a good seal the required quantity of heat must be transferred to the packaging in the required time. A build-up of deposits on the hot sealing heads can result in slower heat transfer and therefore inadequate seals. Cleaning of sealing heads is a dangerous task. The surfaces are hot and are, in some systems, accompanied by sharp cutting blades. Some companies use powerful cleaning chemicals to clear the carbon from the heated surfaces and these chemicals can cause hazards. Another common method for keeping the sealing surfaces clean is the use of wire brushes and metal scrapers and abrasives. The physical cleaning of the sealing surface this way can cause shape changes in the surface and as a result the seal quality can be impacted in the long term.

5.12 Conclusions

This study, though limited in size, is the first attempt to measure the issue of seal integrity with regard to its impact on waste in the food industry.

The cooperation of food manufacturers allowed the study to be shaped to include a broad range of factory types, product sectors and packaging formats.

Seal integrity issues appear to be contributing 180 000 tonnes of waste to landfill each year from the food industry, with perhaps a further 666 000 tonnes occurring in the supply chain post factory with retailers and consumers also generating waste.

The majority of faulty packs appear to be caused by food entrapment in the seal area.

The best practice that has emerged from the study is as follows.

5.12.1 Robust filling methods

Robust methods of filling packs without the possibility of seal area contamination are essential if the largest cause of failure is to be minimised. This is achieved in several ways by different machine manufacturers. Masking systems for tray sealers and horizontal form fill seal systems and pinch bars for vertical form fill seal machines were observed working well during the study. A 'dipping' filler tube for food from an automatic weigher and a dipping nozzle with clean cut-off for sauces were seen as best practice in the manufacture of ready meals. Examples were seen of packs being wiped clean prior to sealing; this, on the whole, was not very successful and still resulted in a large number of rejects and waste. In fact, the wiping operation in some cases actually increased the reject rate for a machine because the seal area contamination was being spread across all seal areas and from pack to pack.

5.12.2 Standard operating procedures

The use of and adherence to standard operating procedures (SOPs) is key in the operation of sealing machines. Sealing temperatures are critical to the strength of a seal, along with the dwell time used. If these parameters are adjusted, the opportunity for seal failure or integrity issues is greatly increased. Machines were observed running 'hot and fast' during the study to increase output. The result was that seals had become overheated and were damaging the film being sealed. The use of new polylactic acid (PLA) packaging was observed in several factories. It was apparent that the PLA materials needed much tighter temperature control, with even small temperature fluctuations causing packs to fail.

5.12.3 Planned preventative maintenance

PPM is essential to the performance of sealing machines. Machinery care by the machine operator was seen as being the most robust form of PPM. At one factory an individual had been selected and trained in the maintenance of all sealing heads for the tray sealers used. The heads were maintained to a very high standard and as a result the sealing performance was among the best observed. Several factories visited had no system for the routine inspection of sealing rubbers on their machines. As a result the rubbers were only being replaced when they started to make leaking packs. In the better performing factories sealing rubbers were replaced on a regular basis by a trained person to ensure they were correctly fitted and in good condition. Several sealing systems were observed that displayed issues of head or jaw alignment. As the parts of the machine came together to make the seal the pressures were not even. As a result uneven seal widths were observed on dye penetration tests and these machines produced a high number of faulty seals. A final area of concern observed during the study was the control of sealing temperature. Temperature sensors were not working correctly on several machines and as a result faulty packs were being manufactured.

5.12.4 Seal testing

Seal testing was mostly carried out using a manual squeeze test. The best practice observed was in a factory using modified atmosphere packing on its products. Each line was equipped with a vacuum chamber machine with a built-in carbon dioxide sensor. Any leak could be detected using this system. The issue was that the system was off line and sample packs were tested every 15 minutes. The best on-line system seen used an ultrasound sensor to check that a small vacuum was present inside the pack. This system checked all packs but was difficult to set up and rejected some good packs. It also relied on a hot fill system to generate the vacuum, so would not be appropriate for all sectors.

5.12.5 Rework of packs

Rework of packs with seal faults can only be achieved if the fault is detected in the high-risk area and the pack is rejected before it passes to the low-risk area. Where this can be done, the waste going to landfill can be reduced substantially. Some of the factories employed a person to check packs, albeit only with a manual squeeze test, immediately after leaving the sealing machine and prior to them passing to the low-risk area. There are two issues with this. The first is that the cost of the checker is high and the motivation difficult to maintain, indeed with some lines running at speeds of over 200 packs per minute one person could not check all packs anyway. The majority of seal faults are caused by food entrapment in the seal and are fairly random in nature – the chance of any system relying on sampling detecting a random event reliably are very low. The second issue is one of time. Most heat seals require a cooling period of several seconds for the packaging materials to bond – testing by squeeze test too soon after sealing could actually make the issue worse.

5.12.6 Packaging materials

While there were some reports during the study of packaging faults being a cause of seal problems this was not observed during the tests. Packaging material variation will be observed more easily in the factories if SOPs and PPM are to a high standard. In the absence of robust management of seal integrity, packaging outside of specification is unlikely to be the cause of most of the landfill generated in sealing operations.

5.13 Moving forward

The study has pointed the way towards some changes that could be implemented to reduce the landfill caused, throughout the supply chain, by faulty heat seals.

- The development of systems to ensure that the seal area is free from contamination prior to sealing.
- The implementation of PPM systems to ensure that the machines are in good condition and used properly, in particular the sealing rubbers on tray sealers, the head/jaw alignment and the control of temperatures.
- The introduction of SOPs with the discipline to operate to the standard required for the machine and product.
- The implementation of seal testing systems to replace the over-reliance on manual squeeze testing. The testing should be of 100% of all output in order to find the random failures as well as those due to the machine or the system.
- Rework should be maximised by testing for seal integrity as early as possible after sealing and while the product remains in the high-risk area of the factory.
- The impact of overheated seals on the ability of the packs to withstand distribution and display needs to be assessed in line with the findings of this study that the 'go to' response for poor sealing is to increase sealing temperature. Packs can appear sealed but after a few hours the structure of the thermoplastics can change and cause a high risk of seal failure in the supply chain.

5.14 Lessons for packaging managers and supervisors

Lessons from the study that will be useful for all people investigating seal integrity and leaking pack issues in their businesses are as follows.

- Packaging and filling equipment should be very well maintained to minimise the chances of seal integrity issues.
- Machinery should be correctly set up with the correct temperatures and dwell times to create a good seal.
- Seal area contamination should be eliminated by design and correct operation of the equipment.
- Seal testing should be standardised and root causes of issues resolved rather than simple corrective actions taken to compensate for the root cause impact.
- Cleaning and inspection of the sealing surfaces should take place on a planned basis.
- A robust seal management system should be introduced to ensure that seal integrity is set and maintained as a high priority for the operation.

CHAPTER 6
Seal testing techniques

6.1 Introduction

Leaking seals are sometimes difficult to spot on high-speed packing lines but they cause waste, quality problems and maybe even food safety issues, so the difficulties must be overcome if the trust of customers and consumers is to be maintained. This chapter will review the full range of seal testing techniques used in packaging operations to check for seal faults. The faults that are the subject of the tests can be grouped into three categories: seal strength, leaking seals and seal appearance.

6.2 Seal strength

A seal can be either too weak or too strong for the function it is designed to carry out.

A seal that is too weak may fail during logistics or display. A seal that is too strong may be difficult for the end user of the pack to open. There is a window of seal strength for most packaging applications that gives the correct balance of seal strength and openability (Fig. 6.1). There are different types of seal strength that need to be examined: peel strength, where the resistance of the seal is measured against a peeling force, and burst strength, where a pressure is placed on the whole package to measure the resistance. Different techniques have been developed to give a reproducible result and so create a seal strength measurement system.

Handbook of Seal Integrity in the Food Industry, First Edition. Michael Dudbridge.

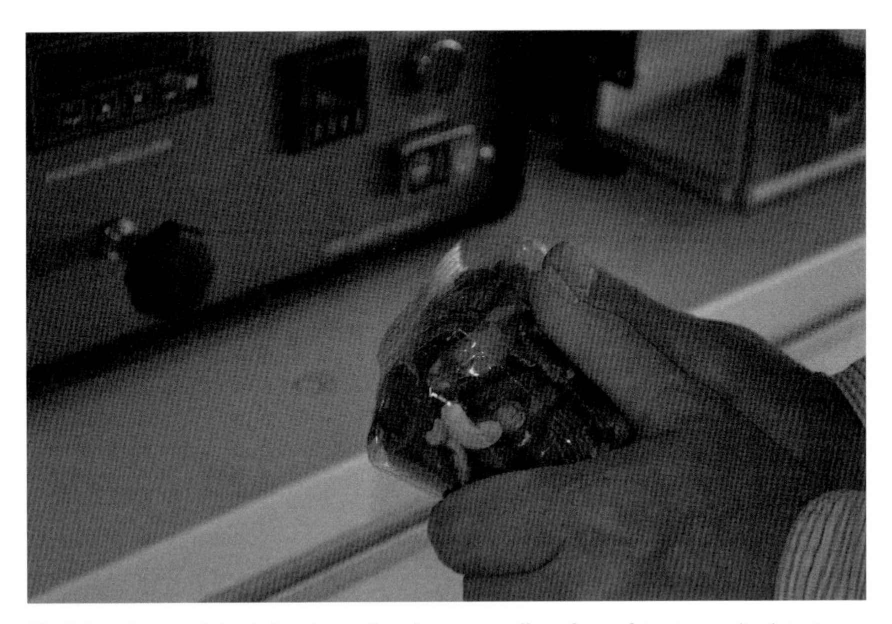

Fig. 6.1 Seal strength is vital and must be adequate to allow the package to survive intact through the rigours of the complete supply chain. There are no physical standards for heat seal strength on flexible and semi-rigid packaging systems apart for the requirement to survive at the lower limit and to peel open (when this is a requirement) at the upper limit. Here we have a typical manual squeeze test to check the seals. This is inadequate at providing the robust data on which good seal integrity management must be based. (Picture by kind permission of RDM Test Equipment Ltd.)

6.3 Leaking seals

A seal is designed to retain the contents of the pack until required. A pack where the seal is incomplete will potentially fail this most crucial of requirements (Fig. 6.2). Different techniques can be used to measure the pack's performance at preventing leaks into or out of the pack.

6.4 Seal appearance

This aspect is especially important in consumer packaging. The seal inspires confidence in the overall integrity of the pack and therefore the quality of the contents. If a seal does not appear to be good then confidence will be lost even if the seal is actually adequate from a functional point of view (Fig. 6.3). Recent innovations have seen a lot of developments in this area where images of a seal can be analysed with the potential to detect anomalies and as a result infer more about the seal than simply its external appearance. Internal seal faults can be recognised from the external appearance. As a result of these developments it is possible to inspect

Fig. 6.2 Finding packs that are leaking is, sadly, a relatively easy task for someone who knows what to look for and where to look in a supermarket. This should not be the case, but it does indicate the size of the leaking seal issues that are present. (Picture by kind permission of RDM Test Equipment Ltd.)

Fig. 6.3 Seal appearance, of itself, is not a technical issue but it does indicate a lack of control in the sealing process and is also an indicator of likely problems with seal integrity or openability. (Picture by kind permission of RDM Test Equipment Ltd.)

all of the packages produced by a sealing system as the tests are non-destructive, and this gives some great advantages over conventional sampling and destructive testing techniques.

6.5 Testing seal strength

Testing of seal strength has traditionally been a destructive test where sealed packaging is sampled from a production line and is tested in a way that means that the package is no longer suitable for sale. This can mean testing the seals until they fail, cutting open the pack to test seals on a test rig or even immersing the packs into water to check for leaks. By definition it is not possible to test all packages in a destructive test, so packs are selected statistically to form a representative sample of the whole of the output. Samples are prepared in a standard way and the test is carried out rendering the samples unsellable (Fig. 6.4).

6.5.1 Burst tests

This is a group of tests where a differential pressure is formed between the inside and the outside of the pack and as a result the pack inflates. The pressure difference is increased until the pack seals fail. The pressure at which the pack fails becomes

Fig. 6.4 Seal strength is a vital issue for consumers and is a big source of consumer complaint. People can injure themselves getting into a package. This is especially true of packages that are opened after microwaving at home. If a seal is too strong on these packages the hot contents can spill out and cause nasty burns. The other aspect of seal strength is where it is too weak, with packs bursting open and causing waste throughout the supply chain.

the test result. The pressure differential can be caused in several ways, of which the following are the most common.

The vacuum test

A sealed pack is placed inside a chamber and a vacuum is drawn (Fig. 6.5a). The reduction of atmospheric pressure outside of the pack causes the air inside the pack to expand and so exert a pressure onto the seals of the pack.

The pressure test

Air pressure is introduced to the inside of a pack through a sealed gland (Fig. 6.5b). As the pressure increases the pack inflates until it bursts and once again the pressure at the point where the seals fail is the result of the test.

There are some common issues with this kind of test for seal strength which have been overcome by different manufacturers of testing machines in different ways.

As the pack inflates there can be a tendency for the pack to distort and so the pressure on the seal areas is not even and predictable. To overcome this it is common to see testing machines that constrain the pack and stop it distorting. This is particularly the case when a sealed tray is being tested.

The rate of inflation needs to be controlled and this will be dependent on several factors. If the vacuum is drawn too quickly there can be a rapid expansion of the air inside the package and this can put additional stress into the seals. If the vacuum is drawn too slowly then there is a chance that a small leak in the pack can protect the pack from bursting and give a false pass of the test. The packaging material of the package is also important. If the material is stretchy, for example a polyethylene film, then this can protect the seals during a test compared a less stretchy film such as a polyethylene terephthalate (PET).

There are international standards that are published by ASTM International (formerly the American Society for Testing and Materials) that set out testing procedures and requirements so that results obtained all over the world can be compared with accuracy. This is especially important in the international world that is the packaging industry.

6.6 ASTM International seal testing standards

Seal testing is standardised by a group of five procedures issued by ASTM International.

6.6.1 Seal strength – ASTM F88 and F2824

Seal strength testing, also known as peel testing, measures the strength of seals made with flexible packaging materials. The seals are peeled apart in a controlled manner and the forces required to peel the seal are measured using strain gauges (Fig. 6.6). The measurement can then be used to evaluate the force required to open a package as well as to check that a seal is consistent along its length. Peel

(a)

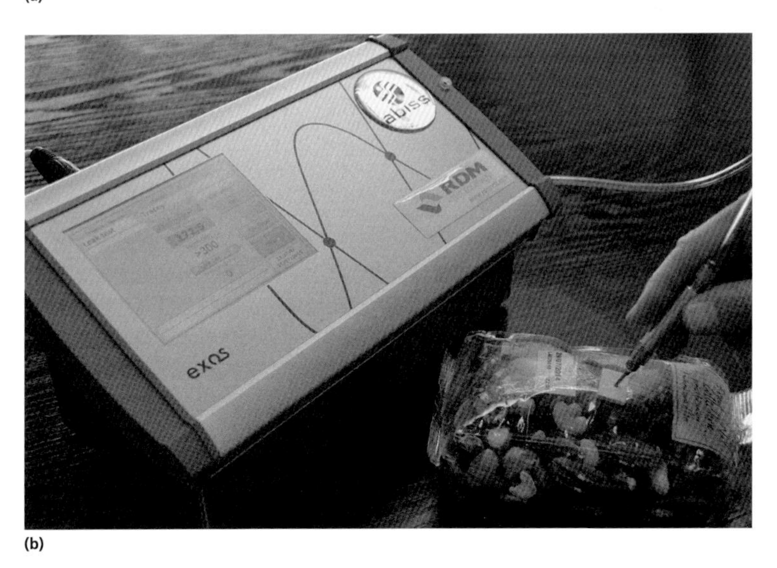

(b)

Fig. 6.5 The creation of a pressure differential between the inside and the outside of a sealed pack can be achieved using a vacuum chamber **(a)** (in this case inside a bubble tank to help visualise any leaks) or pressure from inside via a sealed gland **(b)**. (Pictures by kind permission of RDM Test Equipment Ltd.)

force measurements can be used in factories to ensure that the sealing process is performing as required and to feed information into maintenance and operations systems if the peel force starts to change. Anomalies in seal peeling forces can also indicate pack integrity issues that might be caused by seal area contamination or other causes, so it is a good quality control check for the whole packing process.

Fig. 6.6 This is a typical peel testing system to check the force required to peel a seal apart. The mode of failure of the seal in this test as well as the forces required are important predictors of how the seal will perform in the supply chain. (Picture by kind permission of RDM Test Equipment Ltd.)

6.6.2 Seal strength – ASTM F2824

ASTM F88 carries out its tests on a 1-inch (it is an American standard after all) section across a seal area. The ASTM F2824 method examines the whole seal area in a pack – the whole seal area is peeled and measured. It is usually applied to the peeling of a lid from a rigid or semi-rigid tray.

6.6.3 Burst test – ASTM F1140

The burst test is used to determine overall package strength – usually the weak point of a package is the seal area but this is not always the case. A burst test requires that the pack is pressurised and that the pressure is increased until the package bursts. The data recorded from a burst test is usually the pressure at which the package bursts and also the position on the package of the failure, as this indicates the weakest part of the pack.

6.6.4 Creep test – ASTM F2054

The creep test is a true seal strength test which gives a pass/fail result or a survival time result after putting a package through a standard set of test conditions. The package is pressurised to 80% of the typical burst test result. Once the pack is pressurised, it is observed to see if the seals are being forced apart by the internal pressure of the pack. Typically, a standard survival time will be set which will then be used to judge if the seals could withstand sustained forces trying to pull them apart. Once again, the position of the failure will be noted as an indication of the weakest point of the package.

6.6.5 Vacuum dye or dye penetration test – ASTM D3078

This test uses a high-intensity dye to identify seal area problems (Fig. 6.7). The dye is drawn into seal faults by capillary action when it is placed inside an empty pack or sometimes a vacuum is used to draw the dye into the package from outside.

Fig. 6.7 A dye penetration test is a good indicator of a faulty seal but can also indicate faults on the sealing machine that have yet to cause a leak but are weakening the seal so that it may not prove adequate during distribution.

The standard methods described by ASTM International are applicable for standard packaging formats but often a package does not lend itself to these standardised methods because of the package design. In these cases a bespoke test is often developed to provide information for the packing factory, but these results are unstandardised so cannot be used between factories. These factory designed tests are useful and important, but care needs to be taken that the design of the test is reproducible by all people who will carry it out. All people, including all shifts and all departments, need to have a standard operating procedure to ensure that a test result can have maximum impact in the business. An example of poor test design was mentioned in Chapter five. The manual squeeze test is not an example of a test that can generate any kind of data for improving a business performance. False positives and false negatives will occur all of the time, so the advantage of a quick easy test is lost because the results are almost worthless for spotting patterns and trends. An example of a bespoke test that is well defined and universal in a factory would be something like the following test for pouches.

6.7 Pouch integrity testing

Pouches are small and often quite complex in shape and structure. Gussets are often built into pouches to increase their volume capacity and it is these kinds of features that make the standard ASTM tests difficult to carry out. A typical bespoke

Fig. 6.8 Pouch integrity test using two plates to squeeze the pouch in a controlled way. This is a uniform test that would yield useful trend information so that a drop-off in seal performance would be spotted as soon as possible.

test for pouches would be to apply a force to the pouch by squeezing it with a known force between two parallel plates (Fig. 6.8). The force would be applied for a predetermined time in order to stress the pouch more than would normally be experienced during distribution and use. The force to be used can be determined by experience to produce a definite failure in a pouch if there is a low probability it would have failed in distribution. The idea of the parallel plates is to get some uniformity and reproducibility to the test that will allow it to be carried out simply and correctly by a wide variety of people. The plates allow the force applied to be evenly distributed to all of the seals of the pouch and so the test will identify the weakest point on the seal.

All of the ASTM tests above are destructive in nature – for a test to be carried out on a package it has to be destroyed. This means that a low proportion of packs can be tested. The following are some more examples of destructive test designs that, while they are not to an ASTM standard, can provide useful data to the business carrying them out.

6.8 Seal strength by inflation and seal integrity testing (destructive tests)

Seal strength and seal integrity testing are two different types of test, both of which provide important information about a pack's ability to protect its contents from contamination or leakage. Inflation tests, such as burst testing, creep testing, creep-to-failure testing and pressure decay leak testing can be used to examine packages for their seal integrity and also their seal strength.

6.8.1 Burst testing

It can be useful in factory environments to develop local burst tests that can give a machine operator an early indication of any issues in the packing process. These local tests do not need to follow the ASTM standard but need to be robust enough to be repeatable from one day to the next and from one operator to another. A burst test can form part of a quality testing system in your factory and inform the factory about trends in the results that might indicate the need for some corrective action before the packs coming from the production line have seals that are bad enough to become rejects.

Typically, a bust test result will produce a pressure (or vacuum) level at which the failure occurred. This number can be captured and plotted onto a control chart along with a line that indicates the minimum and maximum acceptable readings for the test. If a trend is spotted that sees the result of the tests getting closer and closer to the point where the packs become unacceptable then early corrective action can be taken (for example, cleaning the sealing jaws) to prevent the occurrence of reject packs with out-of-specification seals.

A burst test that involves standing on the pack to see if it can support your weight (believe me I have seen this in a factory!) is not an acceptable test if data is needed and trends analysis is required. The manual squeeze test is another example of seal testing that is often used but has no basis in science to allow analysis on the performance of the packaging machine in making adequate seals.

6.8.2 Creep and creep-to-failure testing

We have already looked at the creep test but here is a non-ASTM version that is often developed locally and provides useful local information for the control of the sealing process. It is a destructive test in that even if a pack survives the test it will have been altered by the rigours of the test and so would not be suitable for sale. In the creep test the package is inflated either by internal pressure or by external vacuum. It is then held for a set time in this inflated state. If the seals survive, that is a pass. If the seals give way, that is a fail. The creep test can be considered as a slow peel test where the thermoplastics that have been joined together during the sealing process are pulled apart by the pressure difference between the inside and the outside of the pack.

One issue with this test is that it does not give a definite number or score as a result. The seal areas are examined after the test to find out the extent of the creep that has occurred and to try to ascertain the quality of the seal from that information. The ASTM method for peel testing looks at just a small section of the seal to achieve a test result, whereas the creep method does examine all the seals on the pack and finds the weakest point.

A development of the creep test is the creep-to-failure test. The pressure (or vacuum) is maintained until the seal actually fails and it is the survival time that gives the result that can be recorded for trend analysis and corrective action. The location of the failure is also more obvious and so the weakest part of the seal can be identified and corrective actions can be more focused.

6.9 Leak detection

There is a whole group of destructive testing methods for the evaluation of leaks in sealed packages. For a leak to occur, there are three factors in play at the same time. These are the size of the hole or channel, the pressure difference on each side of the hole or channel and finally the viscosity of the fluid material flowing through the hole or channel. For example, a hole may not leak if there is no pressure differential to encourage the flow, but remember that the flow of gases through a hole is determined by the partial pressures of the individual gases on each side of the hole. So in a pack of sliced ham with a modified atmosphere inside the pack of 30% carbon dioxide and 70% nitrogen the partial pressures would cause nitrogen to flow into the pack from the 80% nitrogen outside. The carbon dioxide would flow out of the pack towards the very low carbon dioxide levels in the atmosphere and oxygen partial pressures would encourage oxygen molecules to flow into the pack to try to equilibrate the atmosphere inside the pack with that surrounding the pack.

6.9.1 Pressure decay method

Pressure decay testing methods involve the pressurisation of the pack with gas to a level that is around 50 to 70% of the pack's typical burst pressure result. The gas is normally introduced using an injection needle through a self-sealing patch (septum) placed on the pack. When the test pressure is reached, the gas flow into the pack is switched off and the pressure in the pack is monitored for around 10 seconds. If the pressure inside the pack decays (reduces) this is an indication that the gas is escaping from the pack. The leak could be from the seals or it could indicate a pinhole in the walls of the pack. The third possible source is from cracks in the packaging material. If the pressure is maintained in the pack to the level set in the specification set up with the test then the pack passes the test. There is, however, the possibility that the pack is designed to leak at a known rate through the use of permeable packaging materials or micro-perforations in the pack. In this case the rate of tolerable decay needs to be set and the actual decay is compared with this to see if the pack meets specification and is acceptable. Permeable packaging materials are often used in products where respiration is going on inside the pack, so are popular in the packing of fresh produce where a controlled level of oxygen flow into the pack helps to extend the shelf life of the product. The final complications of the pressure decay method of testing occur when the packaging materials are slightly elastic in nature and slowly stretch when the gas pressurises the pack. The stretching will make the volume of the pack greater and as a result the pressure inside the pack will drop even if the pack is completely sealed. The same effect will also be seen if the atmospheric conditions change around the pack during the test. Temperature changes of the pack have a big impact on the pressure, as can changes in atmospheric pressure, so the tests need to be carried out in stable conditions. In a food factory the high-risk areas where food is usually sealed into its packaging are usually pressurised above normal atmospheric pressure to reduce contamination, so the simple opening of a door can expose the pack to a change in atmospheric pressure that could lead to a false test result. Data from all seal testing should be used to plot trends in packaging machine performance so that the maximum value can be extracted from the tests in terms of preventing inadequate packs reaching the end user.

6.9.2 Vacuum decay method

The pack in this method is exposed to a vacuum and then the level of that vacuum is monitored by the test instrument (Fig. 6.9). If the vacuum level decreases it indicates that gas is entering the vacuum area, and the only place that can be

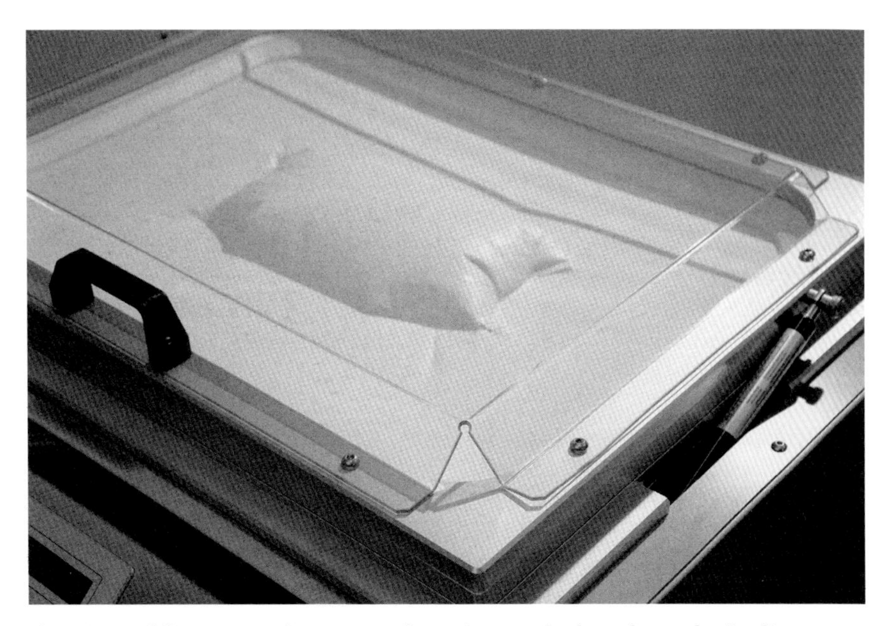

Fig. 6.9 An off-line vacuum decay system for testing samples from the production line. (Picture by kind permission of RDM Test Equipment Ltd.)

coming from is the inside the pack. This method is usually carried out away from the production line on a sample of the output – maybe one pack every 15 minutes from each of the sealing systems.

6.10 Non-destructive tests

All of the above tests can be seen as destructive – the package after testing cannot be sold as it has been damaged in some way by the test. This means that the tested packs are just a small sample of the full production and there is a high risk that a faulty pack will be missed and will end up in the hands of the end user. Ideally, non-destructive testing would be used so that all packs can be examined. To be cost-effective the testing method must be automatic and fast to keep up with the high packing speeds often used.

6.11 Shop floor detection of leaking packs

It is essential that methods of leak detection can be carried out as close to the packing process as possible to facilitate the test and to ensure that results can be acted upon very quickly. Shop floor tests have to be just as rigorous as laboratory-based tests if useful results are to be obtained, so tests need to be simple to carry out and accurate enough to provide the information needed for decisions to be taken about the running condition of the packing line. The issue with most tests is that they are off line and require the destructive testing of packages to get a result. Recent advancement in sensing and computer processing speeds has allowed the development of on-line inspection systems that are capable of measuring seal integrity in a non-destructive way (though any faulty packs detected by these systems are obviously rejected). The ability to test every pack, non-destructively, means that sampling errors are eliminated, providing every customer and consumer with packs deemed acceptable at the point of testing. This is a big improvement on systems where only a few packs are sampled and considered to be representative of every pack being produced. This is very useful when the cause of a leak is a random event – and we have seen in Chapter 5 that most leaking packs are caused by seal area contamination, which is essentially a random event.

There are many types of system developed with lots of variation but they conform to six main groups.

1 The mechanical squeeze test
2 The trace gas detection methods
3 Computer vision systems
4 Vacuum decay systems
5 Vibration and deceleration analysis
6 Ultrasonic systems

6.11.1 The mechanical squeeze test

This method has been developed by several companies in machines that can be fitted into a production line to test the packs leaving a sealing system. The final design depends on the type of pack to be tested, but the basic principles are the same in each case.

The packs are let into the machine and they are squeezed using a mechanical device (it could be a small plate or a running conveyor system). Pressure is applied to the package and then time is allowed for the pressurised pack to stabilise – the pack may change shape slightly or the packing materials may stretch slightly. Once the pack is stabilised, a linear encoder is used to detect movement in the pack resulting from the pressure that is being applied. If the movement is greater than would be expected and acceptable the pack is rejected as it is assumed that the movement is as a result of a leaking pack. These systems are a direct interpretation of what is happening with a manual squeeze test on a pack – except that the sensors and control in the machine mean that small leaks can easily be detected, whereas a manual test would find this impossible.

There are some special cases with a mechanical squeeze test that need to be accommodated into the design of the machine. If a sealed tray is to be tested in this way then the rigidity of the side walls of the tray needs to be taken into account. If the downward pressure on the pack is applied to the top film and to the side walls then the test will become inaccurate as the side walls will resist the movement even if the pack is leaking. This can be overcome in a couple of ways. Either the plate pressing on the pack to increase the pressure must always be smaller than the size of the tray or the tray must be packed with a slight overpressure so that the top film in convex and can be squeezed without the side walls interfering. Packing a tray with a small amount of overpressure is very easy to do on a modern tray sealing machine carrying out modified atmosphere packing operations.

6.11.2 The trace gas detection methods

This is a group of methods that can be used to detect leaks and is commonly used in factories where modified atmosphere packaging is used so that the trace gas can easily be introduced into the pack. The prerequisites for this system are that a gas must be introduced into the pack before sealing that can be easily detected if the pack were to be leaking. The leak of the tracer gas is then detected by a sensor that is specifically designed to detect the presence of that gas. The most common gas used in packing systems is carbon dioxide. It is present at quite a high level (around 20 to 40%) in a lot of modified atmospheres anyway and is not too common in the general atmosphere (around 0.5%), so rises in carbon dioxide inside the detection machine will be a good indicator of a leaking pack. As with all methods of leak detection there needs to be the presence of a pressure differential as well as a route for the gas to leak through for a leak to be detected reliably, so in the case of trace gas detectors the pressure differential is usually provided by the squeezing of the pack or by the application of a vacuum. The application of the vacuum does take a lot of time, with cycle times on that design of machine being around 15 seconds, so on high-speed packing lines the product is often tested in batches rather than individual packs. If a leak is detected it is known to be one of, say, 15 or 20 packs, which then have to be individually checked to find the culprit. Trace gases other than carbon dioxide are sometimes used and are selected depending on the product being packed as well as the cost of the gas and the sensors.

6.11.3 Computer vision systems

There has been a lot of work carried out on the use of vision systems to examine the seal areas of packages to check for anomalies in the seals (Fig. 6.10). There is no way of a vision system being able to identify a leaking pack but it can detect a change in the seal area that would strongly suggest that a leak is likely. The recent advances in computer vision have allowed images of seals to be analysed more quickly, and the advances in image pixel count have greatly improved the resolution of the images being analysed. The combination of these has meant that a computer vision system can spot an anomaly in a seal far better than a human can, and so the use of vision systems in seal inspection is increasing. The final area of vision systems that gives them advantages over human inspection is the use of different wavelengths of light. A human inspection can only take place using the visible spectrum, but a vision system sensor can be sensitive to a much wider range of light wavelengths. Vision systems using thermal imaging with wavelengths in the infrared range and those using ultraviolet wavelengths can help spot issues

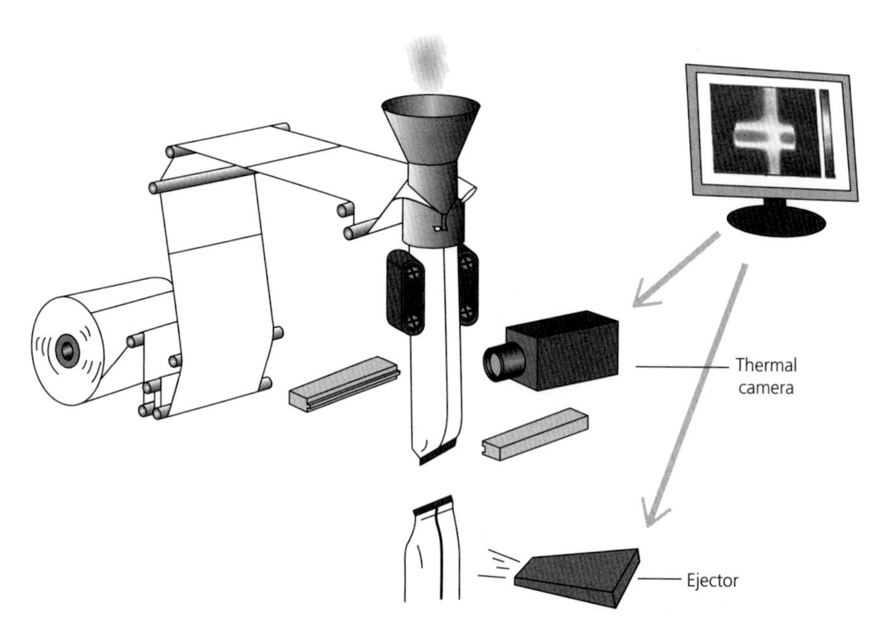

Fig. 6.10 Vision systems allow detailed computer analysis of seals. Computer processing speeds and resolutions are improving all the time and these systems can spot problems that a human would find it difficult to see. (Picture by kind permission of RDM Test Equipment Ltd.)

that are invisible under normal light. Vision systems can also use techniques such as polarised light and laser scatter to help highlight issues with seals.

6.11.4 Vacuum decay systems

We have already seen the off-line pressure and vacuum decay methods that subject a pack to a pressure or vacuum of a known level and then look for a decay in that pressure or vacuum to indicate a leaking pack. On-line inspection systems have also been developed using the vacuum method as it is non-destructive and capable of being automated. A pack is automatically placed into a chamber, the chamber is sealed and a vacuum is drawn to a known level. If the vacuum decays in the following seconds it indicates that the pack is leaking. These machines are designed to test one pack at a time and sometimes have issues in everyday use where the chamber does not seal correctly and so the leak detected is in the chamber and not the pack. The other issue with this type of test is the way that it handles a gross leaking pack. A pack with a very bad leak will evacuate at the same rate as the chamber around it and so when the decay phase of the test is carried out no further decay is detected and the pack will pass the test. This means that a secondary test method is often required to check for the very badly leaking packs.

6.11.5 Vibration and deceleration analysis

When a seal is made it usually involves the coming together of sealing jaws or the closing of a set of sealing tools. By careful analysis of that process it is possible to detect an anomaly in the machine that would indicate that a faulty seal has been made. Take for example the manufacture of a bag on a vertical form fill seal (VFFS) machine. As the sealing jaws close to make the seal the last few milliseconds of the action can be captured and the deceleration and vibration of the metal of the jaws can be measured. The measured patterns can then be automatically compared to a standard profile for that action and so anomalies can be detected and reject signals can be generated. The seal faults that can be detected in this way include where product has become trapped in the seal area and where the packaging material has creased or folded and could result in a leaking seal on the pack. This is a relatively new area which is being developed currently but shows great promise for the future where this kind of technology can be embedded into the packaging machine (Fig. 6.11).

6.11.6 Ultrasonic systems

This method of examining seals on a pack is divided into two main types. The first is based on the patterns seen when ultrasound is sent through the seal and detected on the other side. Systems of this type are seen to be quite slow as the ultrasound source has to make contact with the packaging material, but they can quite easily be incorporated into a sealing system that has an intermittent motion. The main downside of this system for on-line analysis is that the design of the ultrasound generator has to be straight, so the system can only be used to test one seal at a time. The system has found a use in checking the top seal on a pouch

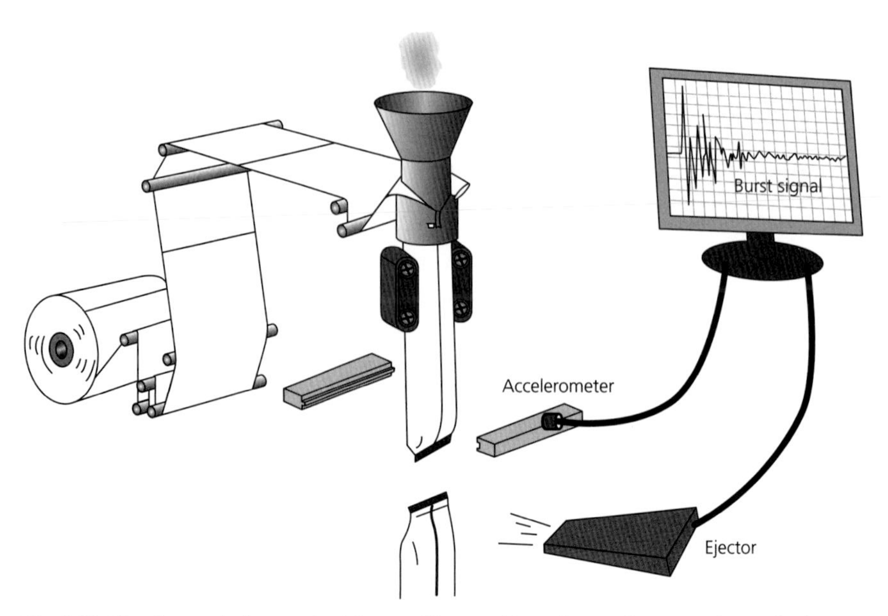

Fig. 6.11 Vibration analysis on a VFFS bag-making machine. The accelerometer is attached to the sealing jaws and can build up a picture of what normal deceleration and vibration looks like. Each operation is then compared to that ideal picture and if it strays outside of the tolerances set the pack can be rejected. (Picture by kind permission of RDM Test Equipment Ltd.)

once it has been filled, but the other issue with this technique is that it checks one seal and not the whole pack, so leaks could be present elsewhere and go undetected.

The second type of ultrasound system for seal detection is based on an ultrasound echo from a 'ping' of ultrasound sent from a generator and picked up by a sensor. The time taken for the return echo as well as the sound frequencies in the echo are used to detect anomalies that fall outside of an acceptable range. These systems need careful calibration but can be accurate if things like filling weight and pack shape are consistent.

CHAPTER 7

Packaging materials and their impact on seal integrity

7.1 Introduction

The sealing conditions required to make a seal during packing operations are dependent on a lot of factors but the most important are those that are dictated by the packaging materials being sealed.

The choice of packaging materials is a complex one where sealing is only one of the considerations. For example, in a consumer product for sale in a retail supermarket the aim is to attract the attention of the shopper so they select your product over a competitor product – so shelf appeal is very important. All of the shelf appeal of a packaged product is imparted by the packaging rather than by the product itself, so considerations of quality of print, the feel of the pack, the quality image of the pack, and the method of display all become very important in the decision-maker's mind. Decisions are also made around the requirements of the product. Does the product need physical protection from damage? How about moisture – does the product need a high moisture barrier? And oxygen and sunlight – does the product require protection from these factors to be safe and of high quality for the end user. Finally, consideration will have to be given to any post-pack processing that occurs, including chilling and freezing as well as post-pack heat treatment processes. What about the consumer – does the package have to be microwave reheated? And does the pack have to be easy to open? Once the pack has been designed, it is then up to the packaging technologist to make it work in the packing process by considering all of the product requirements from the marketing department as well as those from the technical department. Considering all of the requirements will involve testing and there will sometimes need to be compromise on some of them as the final consideration will be the cost of the pack.

Handbook of Seal Integrity in the Food Industry, First Edition. Michael Dudbridge.

I believe that the section above indicates the complexity of the decisions that have to be made around packaging materials, and it will be no surprise that a packaging industry has grown up to be able to meet all of the packaging requirements above. Packs are often designed with some compromises. The compromises made in the design and formulation of packaging materials concern the balance that has to be struck between performance and cost. So packaging materials are often designed to perform well but with the realisation that packaging costs are a substantial part of the overall cost of a food product and that sometimes it is impossible to justify increased cost for a marginal increase in performance. The relationship between cost and performance is, of course, always changing and the position must be closely monitored to ensure the best balance is maintained to deliver value to the consumer. For this chapter we are only going to consider the thermoplastic properties that are important from a seal integrity point of view. There are lots of sources of information for the wider considerations of packaging systems, so we can focus here on the properties that might be important in setting up and running a sealing system.

7.2 What is a thermoplastic?

The accepted definition of a thermoplastic is a material that has physical properties that change as the material temperature changes. As the material becomes hotter it becomes more flexible and as the material cools it becomes more rigid. These are useful properties in a packaging material as it allows packs to be shaped and trays to be formed, but it also allows seals to be made very predictably if the thermoplastic materials are heated and cooled to very tightly defined parameters.

Thermoplastics tend to be made up of long-chain molecules and as such do not have a defined melting point. Instead a thermoplastic material will pass through a glass transition as its temperature increases. So a thermoplastic material simply becomes 'more rubbery' as its temperature increases and will gradually become a viscous liquid. As the temperature reduces the thermoplastic will take the reverse path to become a solid again. It is this behaviour that allows us to exploit the viscosity changes and create seals.

Different thermoplastics have different characteristics and so different packaging materials can have different properties with respect to their behaviour while sealing. It is also possible to manipulate the characteristics of a thermoplastic polymer by the addition of other polymers to the material. These are called co-polymers and are small additions to the mix to incorporate special requirements into the final packaging.

7.2.1 Molecular structure of thermoplastics

Different polymers have different chemical formulae and as a result generate different levels of bonding between the molecules in a material. This intermolecular bonding has a big impact on the behaviour of the polymer when it is subjected to changes in temperature. Some thermoplastic polymers have quite random structures with very weak intermolecular bonds. These materials are called amorphous (Fig. 7.1). Other thermoplastic polymers are more structured, with areas where the polymers line up to form strong intermolecular bonds. These materials are called semi-crystalline (Fig. 7.2 and Fig. 7.3).

Packaging materials such as polyethylene, polypropylene and crystalline polyethylene terephthalate (CPET) are commonly semi-crystalline in structure.

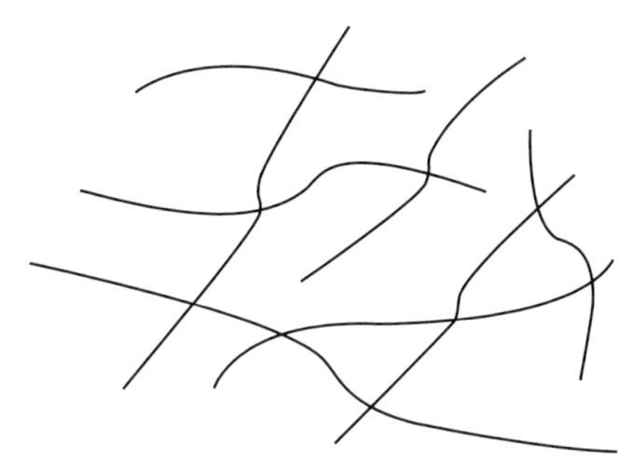

Fig. 7.1 A diagram to illustrate the molecular structure of an amorphous thermoplastic. Note that the molecules have very little order and structure and are essentially a random placement. This structure gives very good light transmission and so a material such as APET can take on a glass-like appearance. The structure also imparts the thermoplastic's reaction to rising temperatures. There is no melting point but instead an increase in the flexibility of the material as its temperature rises.

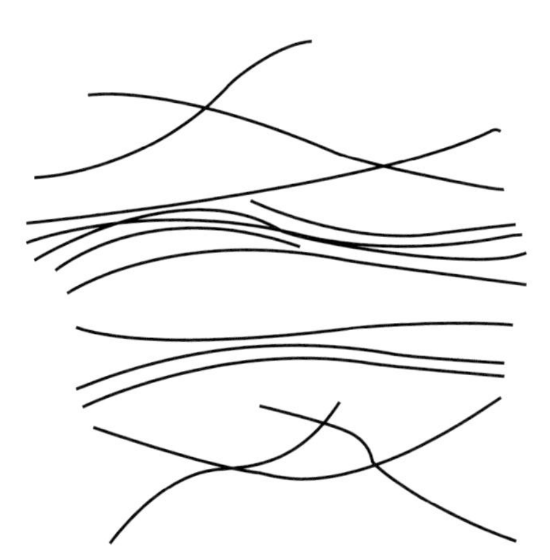

Fig. 7.2 A diagram to illustrate the molecular structure of a semi-crystalline thermoplastic. The material is a mixture of amorphous and crystalline regions and this gives rise to the physical properties of the material. As its temperature increases the amorphous regions become more rubbery but when a critical temperature is reached the intermolecular bonds in the crystalline regions break down and the thermoplastic becomes more liquid in its behaviours.

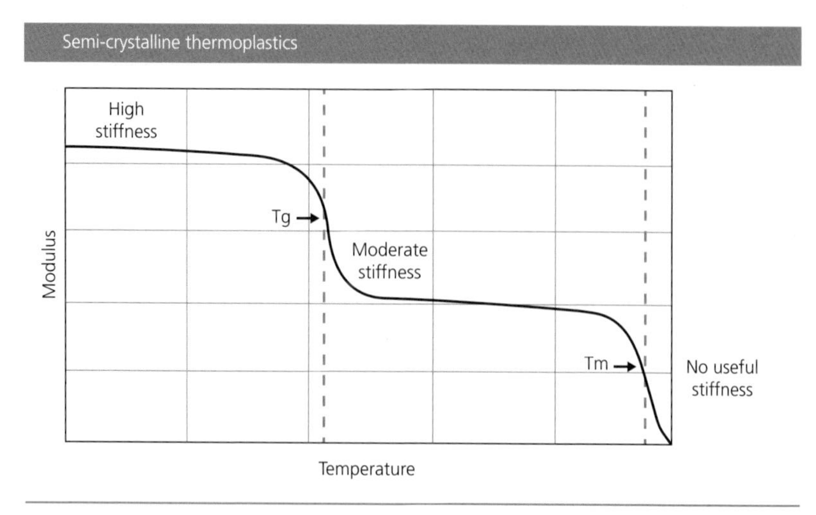

Fig. 7.3 A chart to illustrate the reaction of semi-crystalline thermoplastics as their temperature increases. You can see the glass transition point (Tg) but also a melting point (Tm) where the thermoplastic becomes liquid because the crystalline structures have broken down. As shown in Table 7.1, these occur at different temperatures depending on the thermoplastic being considered. With an understanding of this process it becomes clear that the 'go to' reaction of a packing machine operator, to increase the sealing temperatures when there are sealing problems, might not be the best corrective action. Understanding the way that different thermoplastics react to temperature will help in the diagnosis and correction of sealing faults, but it is important to first understand what materials are present in the packaging materials.

Table 7.1 Glass transition and melting temperatures for a range of semi-crystalline thermoplastics.

Semi-crystalline thermoplastic	Tg (glass transition) (°C)	Tm (melting point) (°C)
Low-density polyethylene (LDPE)	−100	95
High-density polyethylene (HDPE)	−115	100
Polypropylene (PP)	−18	165
Crystalline polyethylene terephthalate (CPET)	76	255
Polyamide (nylon)	47	205

The only common amorphous thermoplastics used in food packaging are polystyrene and amorphous polyethylene terephthalate (APET), where the clarity of an APET package gives it a glass-like appearance combined with great strength. The rigidity of polystyrene makes it easy to use for single-serve pots. The characteristics of an amorphous thermoplastic make it very easy to mould into shapes, which is why APET is used for thermoformed drinks bottles and polystyrene is commonly used for thermoformed yoghurt pots (Fig. 7.4).

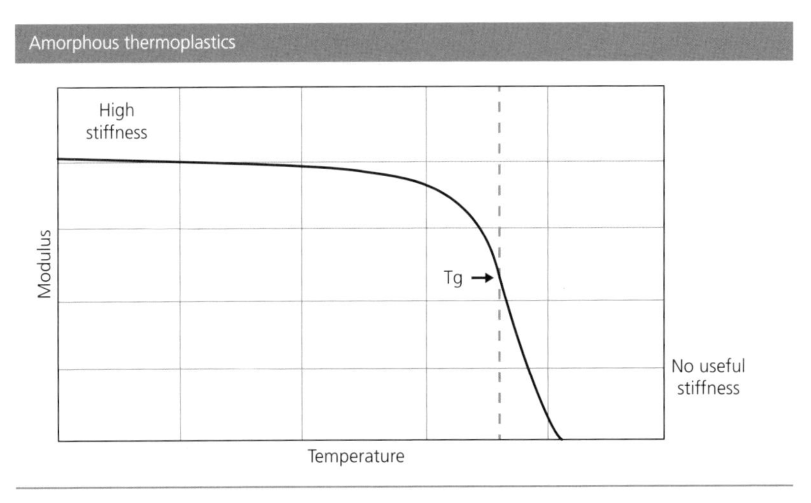

Fig. 7.4 A chart to illustrate the reaction of an amorphous thermoplastic to increasing temperatures. Note that as the temperature increases nothing much happens but then at a critical point called the glass transition point (Tg) the intermolecular bonding breaks and the molecules start to move apart from each other. After that the material simply becomes more and more flexible while remaining a solid material. The glass transition temperature occurs at a different point for different amorphous thermoplastics, so for APET it is 72°C and for polystyrene it is around 100°C. It is for this reason that cups made of polystyrene are not suitable for hot drinks made with water straight from the kettle.

7.3 Commonly used sealable thermoplastics

There are many types of thermoplastic polymers that are used to create seals in packages. Either the package is made entirely of the polymer or the polymer can be used as a layer within a laminated packaging material.

7.3.1 High-density polyethylene (HDPE)

This is the simplest polymer in terms of chemical structure, with a simple carbon backbone with hydrogen atoms attached (Fig. 7.5). The material is classified as linear as it has limited side-chain branching. Because of its simple structure it is possible for the polymer molecules to align and so the structure can be quite crystalline. The material has a good moisture barrier and is commonly used in rigid structures such as bottles and trays. It has a sealing initiation temperature of around 135°C.

7.3.2 Low-density polyethylene (LDPE)

This used to be the most commonly used material for heat sealing. The basic chemical structure is the same as that for HDPE but more side-chain branching has been introduced in the manufacturing process (Fig. 7.6). This has the effect of lowering the level of crystalline structures, lowering the density and also

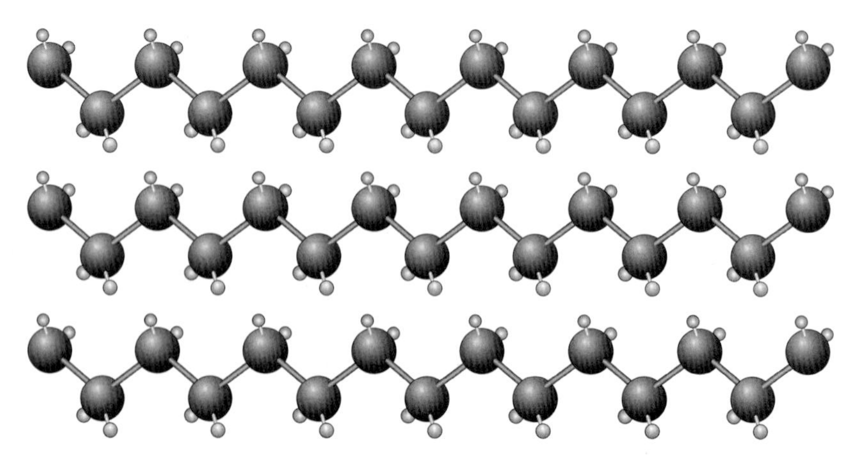

Fig. 7.5 A diagram of the molecular structure of HDPE. The high density occurs because of the alignment of the molecules and the stronger intermolecular bonding that occurs.

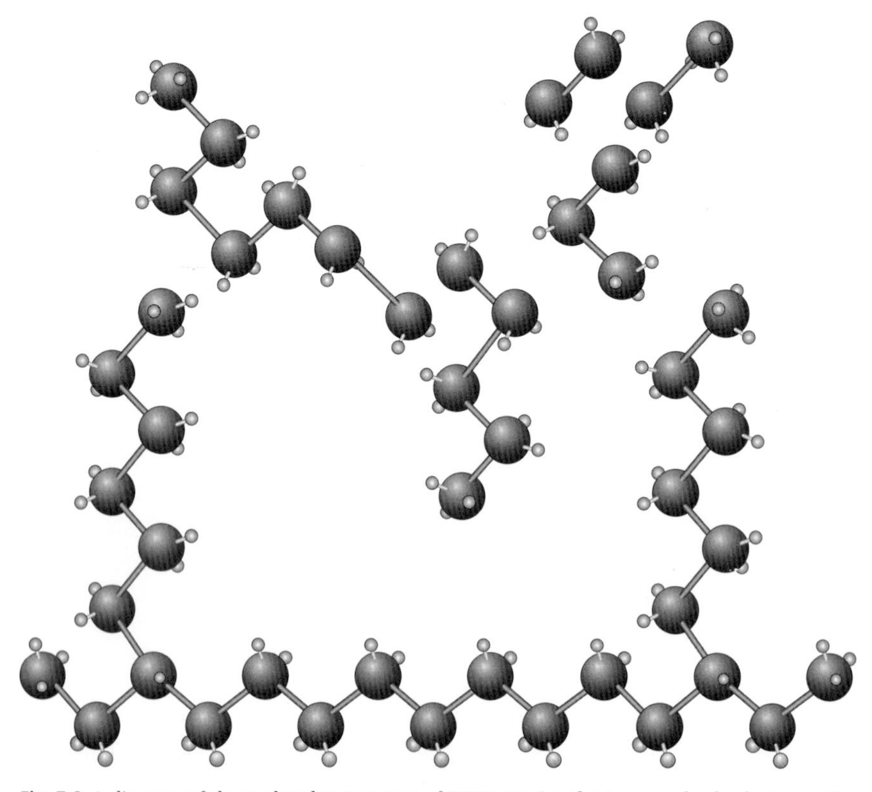

Fig. 7.6 A diagram of the molecular structure of LDPE. Notice the increased side chains on the molecules to reduce the density of the material. The side chains are variable in length and this causes the whole structure of the material to be less strong.

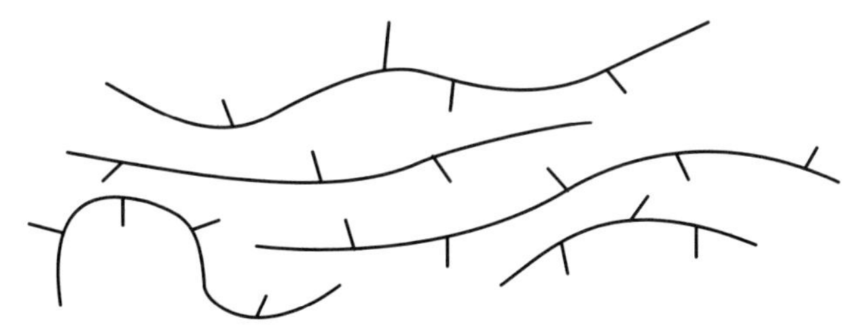

Fig. 7.7 A diagram of the molecular structure of LLDPE. Notice the material has shorter side chains than LDPE.

increasing the flexibility. LDPE has a low seal initiation temperature of between 105 and 110°C and is mainly used in the form of thin films to exploit its strength.

7.3.3 Linear low-density polyethylene (LLDPE)

This has now overtaken LDPE as the lowest-cost polymer for packaging materials, as a result of its popularity and the huge volumes that are manufactured. This material is really a co-polymer where butane, hexane or octane monomers are added to control the side-branching and make a more regular structure (Fig. 7.7). This results in a stronger material than LDPE and so film can be made thinner. Typically, an LLDPE material can be sealed at around 125°C.

7.3.4 Polypropylene (PP)

This material is higher cost that LDPE and LLDPE but has advantages in packing operations. PP can be used at higher temperatures, which makes it suitable for high-temperature applications such as 'hot filling' or packs that are processed in a retort. PP is used for cups, bottles and trays and in some cases films too, when the molecules have been aligned by stretching into something called orientated polypropylene (OPP). Typically a PP material can be sealed at 165°C.

7.3.5 Polyvinyl chloride (PVC)

This material is excellent at thermoforming into trays and packaging shapes and is an amorphous polymer with no crystalline structure. Its use as food packaging has been discontinued because of migration into the food materials and problems with ozone depletion if the material is incinerated.

7.3.6 Polyethylene terephthalate (PET)

This polymer is used in three different forms, depending on the requirements of the product. Orientated PET (trade name Mylar™) is used in a thin film form. Amorphous PET (APET) is used for carbonated drinks bottles where its clarity and strength is required. Crystalline PET (CPET) retains its rigidity at higher temperatures and so is used in ready meal production where the tray formed from CPET can be used in a microwave oven or a conventional oven. There are some other variants of PET available, for example rPET, which is a material made up of recycled PET materials. Typically, a CPET material can be sealed at 250°C.

7.4 Co-polymers

There is a range of materials that are added to the materials above to change the chemistry and the structure and so to impart different characteristics to the final packaging materials.

7.4.1 Acid co-polymers

Acid co-polymers such as ethylene acrylic acid (EAA) and ethylene methacrylic acid (EMAA) are added to polyethylene (PE) to improve adhesion to metals surface, so they are often used in laminated packaging where a metal foil or a metalised area is included; for example, ketchup sachets or alcohol towelettes.

7.4.2 Ethylene vinyl acetate

Ethylene vinyl acetate (EVA) can be added to PE as a co-polymer to impart a lower seal initiation temperature and increased ability to seal through contamination. The EVA gives a lower glass transition temperature and so can improve sealing of frozen products.

7.4.3 Ionomers

Ionomers – where an ionic salt is formed from an acid co-polymer – are used to improve the seals of PE packaging by increasing the resistance of the seal to grease migration. Commonly used in the packaging of bacon, this material has a brand name of Surlin™ and was developed by DuPont. Ionomers also have the advantage of good 'hot tack', where the seal strength is high even when the seal is hot. Also, the use of sodium in the ionomer builds an ionic attraction into the seal.

7.5 Eco polymers

Recent developments have brought a new range of polymers to the packaging ranges of the major packaging suppliers. Materials that are derived from sustainable sources have emerged and are being tried by packers. They also have the advantage of being either biodegradable or compostable as a method of disposal, which does have some environmental advantages.

7.5.1 Polylactic acid (PLA)

PLA is a polymer derived from maize starch and can be formed into packaging materials such as trays and films. The heat sealing of this material is very temperature critical, with the operational window being smaller than for conventional packaging. This has caused some issues with implementation.

7.5.2 Polyhydroxyalkanoates (PHA)

This is really a group of polymers that are grown inside bacterial cells and then harvested to be formed into packaging materials. The base raw materials are fats, sugars and starches. The bacteria use PHA to store energy. This process can be used to create packaging materials with a wide variety of properties.

7.6 Adhesive and cohesive sealing

These are terms often used in discussions about sealing which are worthy of some explanation in this section on packaging materials. It is important to note that in most cases it is vital in heat sealing that the same material is used on each side of the seal. So PP should be sealed to PP and PET to PET and so on. This is because of the nature of what a seal is at an atomic level. There are two basic mechanisms operating in a seal area.

7.6.1 Intertwining of the molecules

As the packaging material is heated the energy has the effect of breaking some of the intermolecular forces holding the materials together. So the ends of some molecules start to break free from the bonds holding them. If at the same time the ends of molecules on the other surface break free and the surfaces are touching then it is possible to get the molecule ends to intertwine and so the two surfaces become joined.

7.6.2 Intermolecular forces

As the seal then cools down the energy dissipates and the intermolecular bonds reform, but this time they are able to bridge the gap between the surfaces, adding strength to the intertwining forces of the molecule ends. This style of sealing is called cohesive, where two surfaces made of the same material are joined.

7.6.3 Adhesive sealing

The second type of sealing is called adhesive sealing. This is where a third material is positioned in the seal area and it is this material that forms a seal. There are many different types of adhesive for various applications but the requirements of the pack contents must be taken into consideration when a selection is made.

A popular example is called 'cold seal'. This is a material based on latex rubber compound that is printed onto specially prepared areas of the packaging where a seal is to be made. The preparation allows the latex to adhere to the base material. When the packaging material is used, the latex material has an amount of pressure applied and a seal is made without the use of heat. Cold seal adhesives are really cohesive as they will only bond with the same material and have a very low adhesion to other materials. Cold seal materials are a special type of pressure-sensitive adhesive (PSA). A strong bond is formed at room temperature with very slight pressure. Sometimes called self-seal, or cohesive seals, the cold seal materials are applied to the packaging materials that will bond together.

7.7 Laminated packing materials

We have looked at a range of packaging materials that are sealable. The packaging industry has used this knowledge to design packaging materials that perform well on high-speed packing system by meeting as many of the needs as possible of all of the users of the packaging. It is quite common for different materials to be laminated together to form multilayer packaging materials to exploit all of the characteristics of the various layers. It must always be remembered that heat seals are better made when the materials being sealed together have the same chemistry and structure. So even on a multilayer package the pack needs to be constructed in a way to allow for this to happen. When a simple error is made in the placement of a roll of film onto the packaging machine and what was going to be the topside of the film becomes the underside the result is that the seals can't be made because the wrong surface materials are brought together by the packaging machine.

Multilayer laminates are designed to optimise the performance of a package in terms of sealing and opening but also in terms of package strength and puncture resistance. Migration of water, gasses and even light is also considered for some products. The appearance of the pack and its ability to accept printing inks is also a consideration. The final, and some would say most important aspect, is the cost of the package. If the economics of the product can be made to work, the product stands a chance in a very competitive market. Packaging material costs are a significant part of the overall cost of a product, so a packaging material decision needs to be correct if a product is to be a success in the long term.

Multilayer laminated packaging can have as many as seven or eight different materials sandwiched together. That is an expensive process and that will be reflected in the final cost of the pack, but if the cost is justified in terms of pack performance the decision to adopt very complex packaging materials can be the correct one.

CHAPTER 8
Seal strength

8.1 Introduction

In this chapter we will consider the implications of seal strength and the potential impact of rough handling during distribution, retail and domestic situations. The seal has to be able to withstand normal handling and this can sometimes include rapid changes of air pressure and temperature changes. The strength of a seal is vital in packaged goods but is rarely given enough emphasis during packaging operations to check that all is well. The chapter will also consider the main methods of seal strength measurement.

8.2 Seal strength – a definition

Seal strength is the measure of the force required to pull apart a seal in a packaging system. This simple definition is a good starting point but you will see later in this chapter that in order to measure seal strength it is necessary to define a test very accurately and carry out the test under controlled conditions if the results are to be meaningful.

Some seals are defined as welds, where the strength of the seal is the same as, or even greater than, the packaging material itself. Other seals are designed to be the failure point in a package and as such the seal strength is less than the strength of the packaging material. So the difference here is the difference in the seal strength of a pouch of microwave rice – where the seals are welded to allow the pouch to be processed through a retort – and the seal strength on a snack food pack where the seal is designed to be peeled open in a controlled way by the consumer.

The seal strength must be adequate for the purpose it was designed for and must fall within defined limits to ensure that the pack does its job correctly (Fig. 8.1).

Handbook of Seal Integrity in the Food Industry, First Edition. Michael Dudbridge.

Fig. 8.1 The strength required within a seal depends on the job the seal has to do within the package. There has been a trend in retail packages towards easy-open seals to make the products more convenient for consumers. It is often said that if a simple tin can was invented today it would be laughed at: 'This food package requires a special tool to open it ... and the can becomes hazardous once opened because of the sharp edges ... and you want to use it to package a consumer food product!'

8.3 The implications of distribution, retail and domestic situations

The seal has to perform throughout the supply chain of the product to ensure that the contents of the pack remain secure. There are several considerations during the supply chain that need to be considered if seal strength is going to be well defined to survive the rigours.

Logistics, warehouses and fork lift trucks all put pressure on package seals. Stacking products onto pallets or placing products into green crates for distribution has to be taken into account if the seal strength is going to be designed correctly to survive distribution.

8.3.1 Transit trials and simulations

It is quite common, when there is a new product launch or a major change in the packaging of an item, that trial packs are sent through a transit trial. The packs are well marked and then mixed in with standard product. They are then inspected at numerous points on the distribution chain to see that all is well with the package. Transit simulations can also be carried out using vibration tables where data

Fig. 8.2 A transit trial or a transit simulation can be used to check that the design of a package and its seals is adequate for the expected distribution loads. There should also be a margin of error designed in to ensure that the package can survive the occasional small 'surprise' of an overenthusiastic fork lift driver or the occasional heavy breaking of a large goods vehicle.

collected from the real distribution system is played back through the table to simulate all of the distribution steps. Both of these systems will check the package and its secondary packaging for the effects of shocks and vibration, crushing and tilting and the full range of forces and conditions that the package will have to withstand. The transit trial should be seen as the final validation of a new package and will help avoid disasters when a product is launched, when the distribution system is changed or when the packaging is altered (Fig. 8.2).

8.3.2 The impact of temperature changes

We have seen in previous chapters that the heat sealing of thermoplastics is a complex task and that the materials themselves are designed to change their properties with temperature. It should come as no surprise then that the full implications of temperature need to be considered when a package is designed and the packaging materials are selected. For example, a chilled ready meal in a polyethylene terephthalate tray will spend time at 2°C during distribution and retail display but might well be marked as 'suitable for home freezing' where it could stay for several months at −18°C (Fig. 8.3). The seals on the package have to withstand all of the possible options for the package. In an attempt to minimise logistics costs it has become common for packs intended for ambient distribution to be moved, for at least part of their journey, in a chilled vehicle. This unexpected temperature can have an impact on the seals and should be considered when logistics operations are set up.

Fig. 8.3 The phrase 'suitable for home freezing' has big implications for a food product but also for its packaging and seal integrity. The package and its seals must be capable of containing the product for all of the expected temperatures it will be subjected to. Freezer temperatures of −18°C can change the rigidity of a thermoplastic considerably and this can have a knock-on effect to seal integrity as the material responds to handling.

8.3.3 The impact of pressure changes

There are several ways in which the pressure inside a pack can change during its life.

A package in a cardboard outer stacked on a pallet can have several layers of product stacked on top. If the cardboard outers are stacked correctly that should not cause a problem for the packs inside, but if an outer collapses or the cardboard softens because it is stored in a chilled area then pressure inside the packs can greatly increase and put seal strength to the test.

In the regional distribution centre (RDC), where the outers from the factories are picked and restacked into cages for onward distribution to the retail outlet, the outer is now in an environment with other outers of mixed sizes and shapes. The carefully planned pallet stacking patterns are lost and it is essentially chance if the outer is going to be damaged on this second phase of logistics from the RDC to the retail store.

Mixed pallets and cages of products and transit cages are a major test for seal strength during distribution. Without the pallet stacking patterns often a corrugated outer case is all that is protecting a package from weight loaded above. This can be considerable and needs to be planned for if this is the method of distribution. A seal that is too weak will open in this scenario and 'faulty seals' will be the reason for rejection and claims by the retail store.

8.3.4 Returnable transit containers (RTCs)

RTCs (plastic crates) can solve some problems but cause others when it comes to implications for seal strength. On one hand, the rigid crate can protect the seals from increased pressure due to crushing, but on the other hand, often the product in the crate is not tightly packed so can move during transport (Fig. 8.4). This can damage seals. Incorrectly used and damaged crates can also pass pressure from the crates above directly onto packages, causing the seals to become strained and greatly increasing the potential for creep in the seal.

8.3.5 Logistics route

This can sometimes have an impact on the seals of a package. The obvious example here is when packages are shipped by air freight. The pressure inside an aircraft will be maintained at the equivalent of 8000 ft when the aircraft is cruising at about 35 000 ft. So the pressure inside the aircraft will be around 800 mBar. If a package in the aircraft was sealed at normal atmospheric pressure of around 1000 mBar that gives a pressure differential of 200 mBar. You can see the impact this might have on seals by looking again at Chapter 5, where it can be seen that

Fig. 8.4 The RTC (sometimes still called the BOC tray after the first company to adopt the system for distribution of retail food packs) has benefits in terms of crushing forces on the seals, but they do subject the packages to more movement during distribution. The potential for one pack to damage its neighbour is greatly increased if the RTC is underfilled. One issue with RTC systems is what happens when the stack of baskets tilts. If the devices in the crates that stop a crate nesting into the one below fail, or if they are not correctly engaged, the weight of a complete stack of crates and their contents can be pressing against one pack in the bottom crate. The chances of the seals surviving that are slim.

many packages would burst open with that pressure difference. Pressure fluctuations can also occur in other forms of transport. Simple things like large lorries passing in opposite directions at speed can create a pressure shock wave which can cause pack seals to react. Simple changes in the logistics route can put extra strain onto package seals that will cause problems with seals that were understrength at the start of their journey.

The effects of changing air pressure can have a marked impact on seals in all parts of distribution. Some modified atmosphere packages are sealed with a small overpressure to try to protect the contents of the package. Quite often food products are sealed into their packages in a high-risk environment of a food factory. One precaution of a high-risk area is that it is slightly pressurised with filtered air to reduce contamination risk. When these packs move to a standard environment they will inflate by anything up to 5% and this could be the difference between a seal being strong enough and it failing.

8.3.6 Retail pressure

Damage caused to packs when they are placed on retail display is a common problem. Packs are often stacked one on top of another, with no consideration of the increased pressure on the bottom pack. The advent of shelf ready packaging (SRP) has helped this problem by keeping the packs in their original carton for display with just part of the outer case being removed. The second form of retail pressure is where the packs become pressurised because they have been distorted in shape through physical damage. If the bottom of a ready meal tray gets pushed in after being mishandled, the volume of the pack reduces but the volume of the contents does not. As a consequence, the internal pressure in the pack increases and this can lead to seal failures if the seal strength is insufficient to cope with the increase in pressure. This is why if the side of a yoghurt pot gets squeezed the lid starts to pop off.

The retail environment is a tough one for a seal to cope with. Products at the bottom of a stack can be subjected to pressure over an extended period and this can lead to seal creep and seal failure if the seal is not strong enough. Add to this the chances of being dropped or bashed and it can be seen that this part of the supply chain can be a major cause of seal failure. The design of the seal must be good enough to survive what is anticipated. This becomes an even bigger issue when packaging materials are being reduced in weight. If the light weighting is not done correctly, pack rigidity can suffer and the pack will distort, potentially putting huge strain on the seal areas of the pack.

8.3.7 Transport home

This is the final area where seals are prone to damage through excess pressure inside the pack. The handling of sealed containers into carrier bags and then into the boot of a car is not the best way to keep them sealed (Fig. 8.5). The placing of heavy items onto lighter ones is an easy way to distort packs and add greatly to the possibility of a seal failure.

Fig. 8.5 The journey home in the boot of the car … enough said!

The seal strength of a product needs to be controlled to ensure that the vast majority of packs can make it through all of the rigours of a supply chain without the seal failing. We will see later in Chapter 9 that the answer to inadequate seal strength is not to weld the seals together so that they will never part. The pack has to be easy to open too.

8.4 Measurement of seal strength

There is an old saying in management that if you want to manage something, first you have to measure it. So, if you want to manage the strength of the seals on your packs, first you have to measure the strength of seals on your packs. The measurement of seal strength is not an easy task and requires the application of some scientific experimentation if it is to be done accurately (Fig. 8.6). Luckily, seal strength is such an important part of packaging that there are many instruments that have been developed to carry out the measurement in a controlled way.

The instruments can be divided up into groups, depending on the particular part of seal strength they are measuring.

8.4.1 Hot tack seal strength

Hot tack seal strength measurements are taken when the packaging material is still hot after sealing. They are a measure of the strength of the seal at elevated temperatures to indicate the performance of the packaging materials in this

Fig. 8.6 The measurement of seal strength is carried out in several ways in factories – some are more scientific than others. Many factories have no way of measuring seal strength apart from the opinion of the quality control operative or the packing machine operator using a manual squeeze test.

important area. Hot tack is important on high-speed filling and sealing machines such as the flow wrapping system or vertical form fill seal (VFFS) system. In both of these systems it is important to gain adequate seal strength as soon as possible after the seal is made. A high hot tack will allow pressure to be put onto the seal soon after sealing and this is required to stop the seal from being damaged when the next product is fed into the packing machine (Fig. 8.7). For example, on a VFFS bag-making system packing 1 kg of potatoes into bags at the rate of 60 packs per minute, the seal will have around half a second to cool down and gain strength before the next 1 kg of potatoes will initially hit the seal and then place pressure on that seal.

There is an ASTM International (formerly the American Society for Testing and Materials) method for hot tack testing of packaging materials (F1921) and this should be followed to ensure that the results obtained can be compared to results in other factories and even other suppliers and other material variations (Fig. 8.8). Knowledge of the hot tack characteristics of your packaging materials can help a great deal in the optimisation of the packing process. The standard method of increasing output of a VFFS machine is to speed it up and then increase the jaw temperatures to be able to make a seal in the reduced time available. Unfortunately, this method has several potential problems. The higher jaw temperatures can give rise to damage of the packaging film molecular structure and this can make the seal area become crinkled and harder than the rest of the

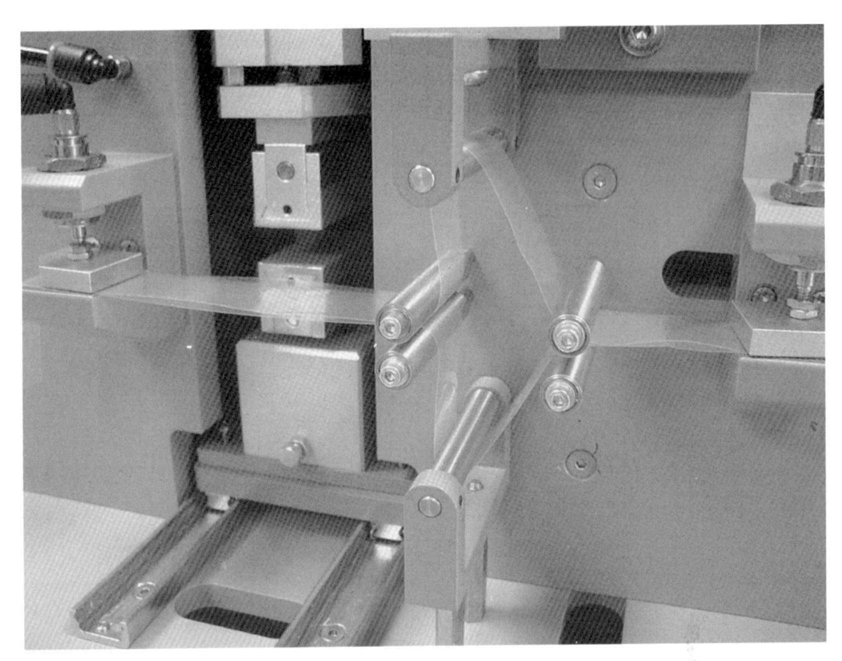

Fig. 8.7 Hot tack is especially important on high-speed packing machines where the seal is going to be put under pressure even before it has cooled and attained its maximum strength. (Picture by kind permission of RDM Test Equipment Ltd.)

Fig. 8.8 A hot tack testing system is something that would normally sit with the packaging supplier rather than the company carrying out the packing operations. It is important though that hot tack strength is specified for each packaging material, and if problems emerge it should be checked. The performance of each consignment of material should be stated on a certificate of conformance supplied with the consignment. This says that the packaging has been tested and that it meets the specification. (Picture by kind permission of RDM Test Equipment Ltd.)

Fig. 8.9 Here we can see an instrument carrying out a test to measure the strength of a section of a seal area. This system does not test complete packs. A burst test would be needed for that type of measurement.

bag. The second potential failing of the 'hot and fast' approach to VFFS set-up is that it takes no account of the hot tack performance of the film being used. Because the seal temperatures are hotter, they take longer to cool down and so longer to reach the temperature where the hot tack seal strength is sufficient to support the loads placed on it. There are solutions to this problem that will allow the speed of the packing operation to be increased. Additional arms can be fitted to the VFFS system that open and close in time with the machine to support the weight of the next kilogram of potatoes while the seal is cooling. This, added to extra cooling of the seal after heating (maybe using an increased air flow), will allow the faster production speeds without hitting problems of product bursting through the bottom of the bags during the sealing operation. The final adjustment that could be made is the timing of the sealing and filling operations. In factories where hot tack seal strength is marginal the optimisation of the timing of the filling will allow for maximum output. The filling operation should be delayed as long as possible once the seal jaws have opened to give the seal a vital few more milliseconds to cool sufficiently that the required seal strength has been obtained.

8.4.2 Burst testing

In this test the package is inflated either by pressurising the inside of the pack or by placing the pack into a vacuum chamber. A number, which corresponds to the seal strength, can be recorded and that can then be used to check if everything is normal on the packing line. This is obviously a destructive test and so is carried out on a small proportion of the packs being made. Typically, as a minimum such a check would take place at the start and end of a shift, at all new product change-overs and after any machine stoppages are carried out. If there have been issues with the seal strength then the frequency of this check can be increased to try to detect problems at the earliest possible time.

8.4.3 The pull test

This is where a sample of the seal area is mounted into the machine that contains strain gauges to measure the force required to pull the seal area apart (Fig. 8.9). This test is used for peel seals as well as welded seals. The mode of failure of the seal is always noted. With welded seals the mode of failure will be the packaging material itself. For the peel seal the mode of failure will be the seal area.

CHAPTER 9
Peelability and openability

9.1 Introduction

One of the major categories of consumer complaint to a retailer concerns how difficult a pack is to open. People see peelable packs as a major advantage. Sometimes these complaints are accompanied with a claim for damages. For example: 'The film lid on my ready meal tray took a lot of force to open and when, eventually, it did give way it required such a force that the contents of the pack were spilled and scalded my hand and splashed down my new designer jeans.'

Considerations of how the consumer will get to the contents of a pack must be at the top of the agenda for packaging designers, packaging materials manufacturers and quality managers, as well as for packaging machine operators. Each has a role to play to ensure that the package behaves as required at the vital stage of opening.

The aim of a peelable film on a pack is to improve the ease of use of the product but without risking the integrity of the pack or its ability to withstand the rigours of logistics and retail display.

Products that are sealed provide great tamper evidence, but if the seal is not peelable then the pack has to be opened with some kind of device. A knife or scissors are often the instrument of choice for consumers, but having to use these devices can pose a danger of injury or damage and is seen as inconvenient. It has been known for frustrated consumers to attack a pack with whatever comes to hand in order to gain access to the contents, and this has resulted in injury and damage to property.

Handbook of Seal Integrity in the Food Industry, First Edition. Michael Dudbridge.

9.2 Peelable films

Peelable films were developed first for use in the medical instrument business to allow easy access to the contents of a sealed pack for people trying to control contamination in operating theatres and medical institutions. It soon became apparent that ease of access was a feature also desired in other markets and this, coupled with the development of polymer technologies, meant that peelable seals could be used in all kinds of packs (Fig. 9.1).

There are three main types of peelable packaging materials. The easy peel feature is designed to operate in one of three different ways.

9.2.1 Delamination

This is where the mode of failure of the seal is a weakness built into a multilayer laminated material that fails as a peeling force is applied to it (Fig. 9.2). If the delamination system of peeling is used in a package it can have a great impact on the operation of the packaging machine. It is usual on tray sealing systems that if a fault occurs – for example the date code has not printed correctly or the top film on a tray is out of register – a tray is opened and then resealed to help reduce waste. With a delamination peeling system the sealing area of the tray is different after peeling so now any seal made on that surface is likely to be impacted. Most likely is that a resealed tray will not have good seal integrity. The only way of overcoming this issue is to rework the pack by putting the product into a

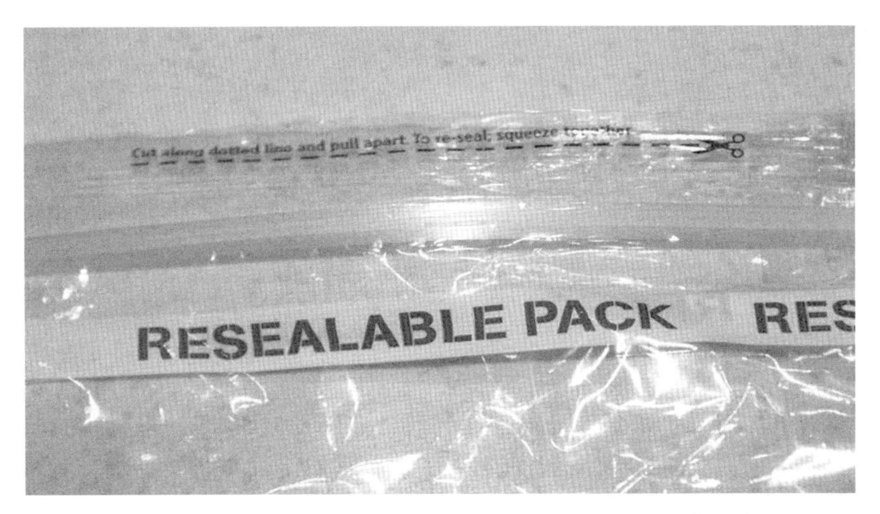

Fig. 9.1 A peelable seal is seen as a desirable feature of consumer packaging. The seal integrity has to be good but the pack also has to be easy to open. These two demands are often in conflict and the packaging industry continues to look for better and better solutions to this problem. Here we see a package that needs to be opened with scissors but is resealable using a zip system. It illustrates the innovation that is occurring but also shows that packaging is often a compromise.

Fig. 9.2 An example of a delamination peel seal where one of the materials that are sealed together delaminates inside its structure to allow the package to be opened. This crisp packet is also showing signs of seal area contamination in the centre of the picture.

completely fresh tray. As this is not always possible or economic to do, the product may have to be scrapped rather than reworked.

This is an advanced form of packaging material design where the packaging film is designed to fail at the junction of two layers. The film is designed to peel by the adhesive layer of the film remaining attached to the base tray or pot and the top layers of the film parting company with the adhesive layer. The performance of delamination peel films does have issues if the conditions that are used to seal them to a pot or a tray are not precisely as required. It is possible, because of the thick layer of adhesive that is used, to generate a fault called 'angel hair' if the sealing temperature is too high or the dwell time is too long. The adhesive becomes more liquid than expected and exudes from the seal area. When the thermoplastic emerges from the outside of the pack it will adhere to the sealing tool and as the tool moves away at the end of the cycle it can pull the exuded thermoplastic into hair-like structures.

9.2.2 Adhesive peel

This is where the layers that were sealed together fail at the junction between them (Fig. 9.3), so the materials retain their original structure and properties after the seal has been peeled open. Because the adhesive peel effectively leaves the materials back in their original state it can leave a pleasing clean appearance. This is not the case with the other two peeling systems where the opened pack will have a residue on the peeled surface. The adhesive peel can be less predictable

Fig. 9.3 The adhesive peel seal is one where the materials peel apart at their original point of joining. Other peeling systems damage one or both of the materials – the adhesive peel is unique in that the materials are left unchanged after the pack is peeled open. This system is used for peelable labels and is also a feature of some types of resealable packaging where the adhesive remains in a workable form and is available to reseal the pack if needed.

than the other two systems in terms of the required peel forces to open the seal, and that is because the peel force is a function of the sealing parameters rather than being determined by the internal structure and chemistry of the packaging materials. Adhesive peeling films are simpler to manufacture and so can be cheaper than the alternatives, but overheating during the sealing process can lead to the pack being effectively welded shut, making it impossible to open without a sharp object.

9.2.3 Cohesive peel

This is where the sealing layers of the packaging material fail internally. This requires some complex chemistry inside the sealing layers to produce a material where the cohesive forces between the molecules are weaker than the adhesive forces in the seal itself. This design of peeling seal is more expensive to produce but has the advantage that the strength of the seal can be greater than the peeling forces needed to open the pack. So a pack can be robust in distribution but still be peelable by the consumer, especially if some easy-peel features are designed into the pack to facilitate the peeling process. Cohesive failure packaging films need to be slightly thicker than the adhesive alternative to allow sufficient thickness in the sealing layer for the failure to occur. The increased thickness can have an impact on sealing times because of the extra time required

Fig. 9.4 The sign of a cohesive peel on a package is the residue that is left behind after the pack has been opened. Here we can see the cohesive peel line on the rim of this tray. The top film has failed internally, allowing the pack to be opened.

for heat to penetrate the film to form the seal in the first place. A cohesive seal is easy to spot as the cohesive failure leaves a residue on the sealing surface (Fig. 9.4). This is sometimes an acceptable compromise when the other factors are taken into consideration. The transfer of material from a top film onto the seal area of a tray can also be used as a quality assurance measure in factories. When a sealed pack is opened the quality of the seal can be examined by looking at the evenness of the seal 'scar' that is left behind. The deposit should be even all around the perimeter of the tray. If it is not it could indicate issues with the sealing machine. One other area to note with cohesive peeling systems is the reduced ability to rework the pack in the event of a rejection, for example if a tray pack is rejected from a production line because it is slightly underweight. The technique of reworking the pack by taking off the top film, adding more material and then sending the tray through the sealing system again may not be successful because of the residue on the seal area.

9.3 Sealing machine set-up for peelable films

In general, peelable films require a more consistent treatment of the thermoplastics in the sealing area to work correctly and so it is best to set up the sealing machine to run at slightly lower temperatures and with a slightly longer dwell time to achieve the heat transfer required. This set-up tends to optimise the hot tack potential of the film being sealed and so it will give a more consistent sealing, and peeling, performance. The tolerance for the set-up on a peelable system is much tighter than that for a welded seal. In a welded seal, the temperatures, dwell times and pressures need to meet a threshold, but above that the seal will still be a weld – except if the sealing conditions are so high as to cause physical damage to the packaging materials. It is a bit of a one-way bet. With peelable seals that is not the case. If the conditions are too severe here, a peelable seal will not be made and a welded seal will occur. This is a particular issue where the 'go to' response of the machine operator to a leaking pack is to turn up the temperature, and this is why, in many factories, the seals are made of peelable materials but they have been taken above the upper limit of the conditions they can tolerate and as a result they are welded shut (Fig. 9.5).

Fig. 9.5 It is possible to convert an expensive peelable package into a welded one simply by operating the sealing machine outside of the acceptable limits. Too high a temperature, dwell time or pressure will destroy the special features in the packaging materials that make it possible to peel a seal made with them.

9.4 The age of the film

This factor is often overlooked when diagnosing sealing faults and can be of critical importance in some packaging systems. The sealing performance of films can change over time, even if all of the other parameters are kept constant. So an older film of the same type will produce weaker seals than a younger version of the same film. The lesson here is to ensure consistent seal performance by rotating stock correctly and ensuring that the principles of 'first in, first out', or FIFO, are observed in packaging stores (Fig. 9.6). It has also been observed that once the seal is made, it will become up to 10% weaker over a period of a few months. This is very important for packages with a long shelf life. The mechanisms at play in film ageing are not fully understood but it is thought that migration of water from

Fig. 9.6 Stock rotation on packaging materials is essential to prevent the build-up of old stock while newer materials are being used. Storage also needs to be in the correct conditions to make sure that the packaging materials operate correctly. This can sometimes be a problem as logistics managers try to reduce transport costs by picking up packaging materials on the return leg for a distribution vehicle. If the distribution vehicle is chilled or frozen, the temperatures and humidity can impact on the packaging even before it has arrived at your site.

the atmosphere into the packaging materials may have an impact, along with structural changes in the polymer thermoplastics.

9.5 The design of the seal area

For a consistent peelable seal it is best to make the seal area as wide as possible. Typically, seal widths on trays and pots are around 3–4 mm and seal widths on snack food bags are made up of five lines of 2 mm each. This wide seal, in both cases, allows the peel resistance per millimetre to be reduced but the overall strength of the seal to be maintained to a level required to withstand the rigours of distribution. Narrower seals of higher peel resistance will be more difficult to open and are more likely to give way rapidly and uncontrollably rather than have a consistent and steady peel resistance. A wider seal width also decreases the opportunity for the creation of channel leaks where an open path can be found between the inside of the pack and the outside.

9.6 The selection of a peelable film

The packaging supply industry is able to develop films with known properties for specific applications, but inevitably specialist peelable films are more expensive than the type of film that is simply welded into position during the heat sealing process. Multilayer films can be designed with peelable systems by the addition of additives and clever manufacturing techniques, and the methods of gaining a peelable seal are now well understood by the packaging companies.

9.7 Product condition when sealing

It is important to take the condition of the contents of the pack into consideration when choosing the packaging materials to obtain a peelable seal. For example, are the products frozen, greasy or hot? If the product is being hot filled into the pack there will be a quantity of steam around the seal area before the seal is made. This high temperature and high humidity situation may have an impact on the desired sealing conditions and may mean that a different peelable film should be specified. Obtaining a reliable seal that peels predictably is not a one-size-fits-all situation.

9.8 Openability

Peelable packaging materials have been developed in direct response to consumer needs for packs that are easy to access. The other set of factors that a forward-thinking packing company needs to consider is other design features that could aid the consumer in getting to the contents of a package with the minimum of fuss. These design features are often incorporated into the package during the packing and sealing operation and vary from quite simple and inexpensive aids to opening all the way to fully featured additions to the package.

9.8.1 Why consider openability?

It is becoming increasingly important to consider the end user in the design of consumer packaging. The average age of consumers is increasing as life expectancy continues to rise (Fig. 9.7). That good news for us all does change things though. As people get older their ability to grip and manipulate packages becomes diminished, and so packages – and especially heat sealed packages – need to be designed to help older people gain access in spite of the potential problems.

Here are some of the features that are routinely incorporated into consumer packaging to aid the opening process without effecting the seal strength or pack integrity.

9.8.2 The tear notch

This is a simple device that gives the consumer a start to tearing the packaging material to gain access to the contents. Typified on a tomato sauce sachet, they work really well providing the packaging material is able to tear. This kind of system is also used on pet food sachets. The notch needs to be applied in such a way that the tear occurs below the seal line of the package. The tear notch is usually applied at the same time as the seal is being made and is the result of a correctly positioned sharp blade in the sealing system (Fig. 9.8).

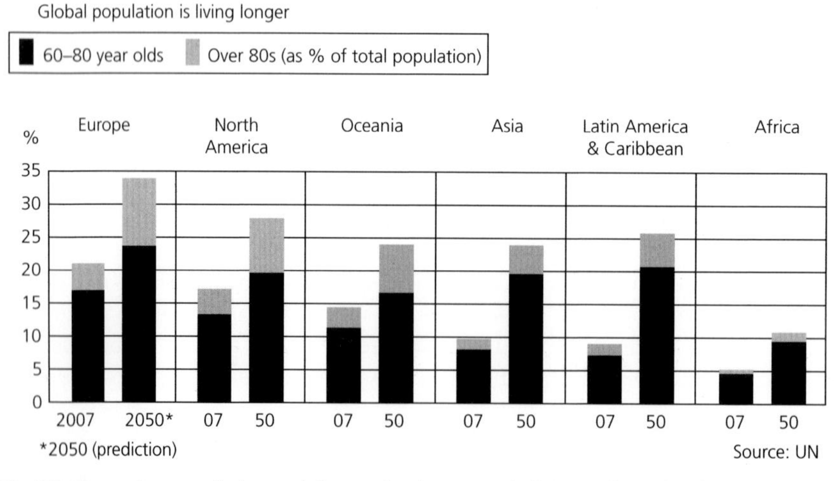

Fig. 9.7 The ageing population and the need to keep people living independently for as long as possible have led to a great deal of emphasis being placed on packaging design by companies keen to serve this growing market. As the population get older the issues of openability of packaging will continue to grow. The chart shows that populations are ageing all over the world, so packaging systems will become increasingly important.

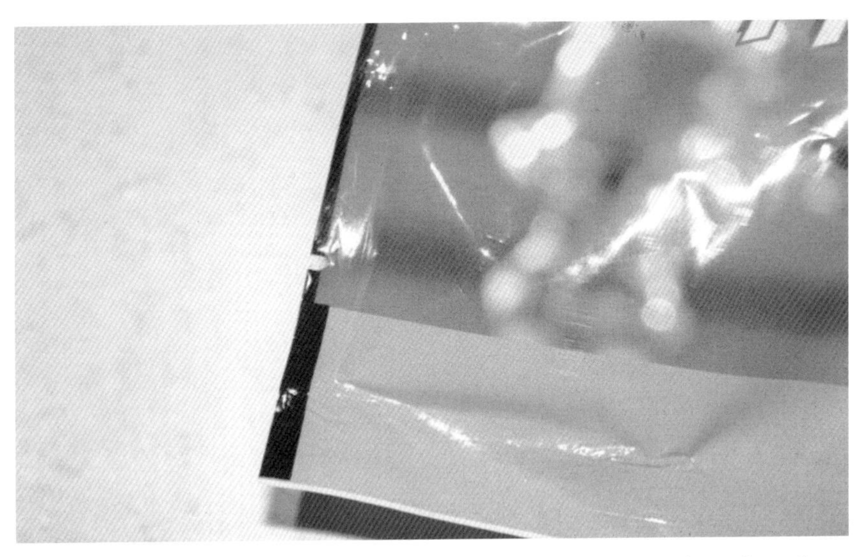

Fig. 9.8 The tear notch is perhaps the simplest of easy-open features but it does rely on the material being torn being linear in nature so that the tear propagates correctly through the material to achieve a clean opening rather than making a random path through the material.

9.8.3 The peel tab

This is the system developed to aid the opening of a tray or a pot container (Fig. 9.9). Typically used on such products as yoghurt and ready meals, it is present on a large number of packs that are designed to be opened by peeling. The peel tab is usually positioned in the corner of a pack if possible (that is difficult on a round yoghurt pot!) and this is done to aid the peeling of the seal. A corner means that a relatively small part of the seal is exposed to the opening force at the start of the operation. This means that a relatively large force is applied to a small area of seal, and that leads to good peel initiation. Once the peel has started, the areas being peeled are simply the width of the seal times two. So the top film on the tray or the pot can be successfully removed without too much drama. There have been some innovations around the peel tab that have helped improve performance in certain applications. Recent trends to help older consumers have seen peel tabs increase in size and the profile of the seal change to make an easier lead in to the peel operation.

9.8.4 Perforations and lines of weakness

This is where the packaging material has a line of weakness designed into it to make it easy to open (Fig. 9.10). Obviously this aid to opening should not compromise the integrity of the pack or allow leakage where this is critical. Advanced systems have allowed partial perforations where the weakness is in the structural

Fig. 9.9 The peel tab is a common feature on ready meal trays and yoghurt pots and helps the consumer open the pack by providing a grip position.

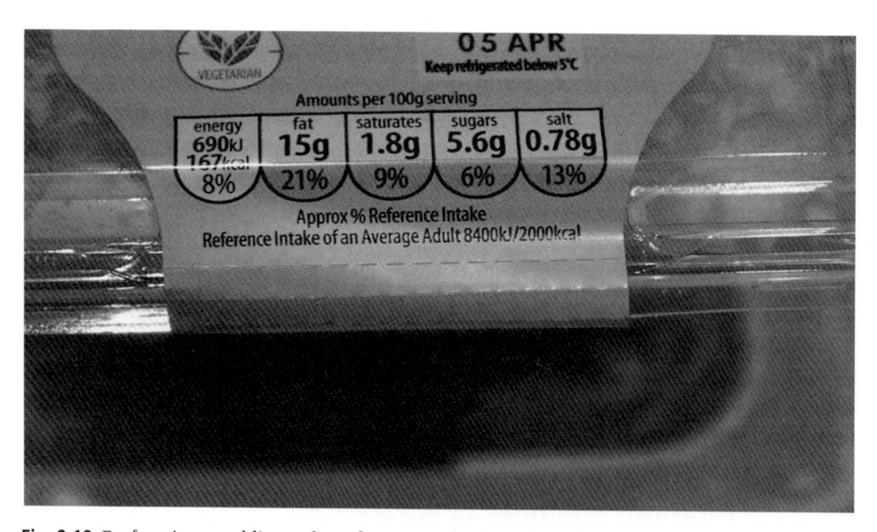

Fig. 9.10 Perforations and lines of weakness can lead to seal integrity issues so cannot be applied to every package, but where they are used they provide a useful aid to opening.

Fig. 9.11 The rip strip is often used on outer film wraps on products. One obvious example is the foil cap for a bottle of wine. Another would be on the film overwrap for a box of tea bags. Here is a picture of a rip strip on a packet of biscuits that otherwise would be difficult to open without the use of a knife or other tool.

part of the packaging material but the barrier to moisture, oxygen and bacteria is maintained. Modern sandwich packaging is made from paperboard lined with a polyethylene sealing and moisture layer. The card is often perforated to allow the pack to be opened easily.

9.8.5 Rip strip

This is where an additional piece of material is attached to the packaging of a product (Fig. 9.11). The additional material is designed to have high tensile strength and is stronger than the main material of the package. This technique is often used on film overwraps of cardboard cartons such as those used for tea bags and cigarettes. The film overwrap on these products is sealed with a welded seal that would be impossible to peel open and so the rip strip allows easy access to the pack for the consumer while providing tamper evidence and barrier properties to the product.

9.8.6 Easy-open lids

We look later at the seals and design features of rigid packaging systems such as bottles and cans, but in a section on openability it would wrong to not include some of the great design work that has gone into making rigid packaging more easy to open. Ring-pull cans are an obvious example where a line of weakness is

Fig. 9.12 The ring-pull can of beans was seen just a few years ago as a new innovation but has now become common in the market as manufacturers try to differentiate their premium beans from the low-cost versions. This is an example of where a packaging innovation around ease of use has helped to create additional value in a product.

designed into a can and lever mechanics built in to break that line of weakness and facilitate the opening of the container (Fig. 9.12). Ring-pulls were first developed for soft drinks but now they are becoming more common in situations where the whole end of a can is removed rather than just a hole being made in it for liquid products. Other innovations have occurred in the bottled drinks market where twist-off crowns have reduced the need for bottle openers.

CHAPTER 10
Resealing a pack

10.1 Introduction

Trends in consumer behaviour have led the packaging industry to innovate with packaging design to increase its ease of use and to improve its performance. One trend that has developed has been in the area of 'resealable' packs. This trend is particularly common in food products to help the consumer see packs as multi-use and so keep food in good condition. A resealable pack helps with the control of portion size as only the required quantity needs to be taken, with the remainder being easily handled by the user and kept in as good a condition as possible, bearing in mind that the seal on the pack has been broken.

10.2 Is it resealed?

Once opened, a resealable pack has lost its integrity and the original 'use by' date on the pack needs now to be ignored. Instead, the storage instructions on the pack should be followed. Often on resealable packs there is an instruction to 'use within 2 days after opening' and this has been calculated to try to protect the consumer provided all of the storage advice is followed. On sauces with a long shelf life that are normally stored at ambient temperatures the storage instructions state that they have to be refrigerated once opened – this is to ensure the food safety of the product and to slow down its deterioration. This emphasises that once the original seal is broken, the resealing feature of the packaging is really just a closure of the pack to prevent foreign body contamination or spillage (Fig. 10.1). The preserving role of the seal is lost even when it is remade. One possible exception to this is where the packaging is providing a moisture barrier for the product, such as with biscuits and breakfast cereals. In these packages the remaking of the seal will be

Handbook of Seal Integrity in the Food Industry, First Edition. Michael Dudbridge.

Fig. 10.1 The resealable pack comes in many forms but for most the pack never regains its full performance after initial opening and the reseal should be considered a closure rather than a seal that is capable of preserving the product.

capable of interrupting the flow of moisture to the product and so could be seen as having a preserving effect.

10.3 Types of resealable packs

In most cases, use of the term 'resealable' is inaccurate. The original seal is broken and is replaced by a closure system that performs in a very different way from the original seal. The resealing feature comes in various forms, depending on the type of pack and the requirements of the product.

10.3.1 The clip-on lid

The clip-on lid use for margarine tubs is perhaps the oldest and simplest type of reclosing device. It protects the product from physical contamination but has very few other features and benefits. This system has been used for a wide variety of other products where multi-use is a requirement. Large tubs of yoghurt and coleslaw are other examples of the use of this technique (Fig. 10.2). The clip-on lid is not sufficient on its own to protect the product throughout the supply chain. Other devices need to be used, such as a heat sealed inner lid or a simple paper barrier in the case of margarine. There is an interesting variation on the clip-on lid called a push lid. A clip-on lid is made by a process of thermoforming, whereas a push lid, often found on containers of gravy granules and drinking chocolate, is formed

Fig. 10.2 The clip-on lid is one example of a popular reclosing system for packages. It is rarely the main functional seal in a package and is intended for multi-use packages, being commonly used on large pots of yoghurt and coleslaw. An exception appears to be yellow spreads, which travel through the whole of their distribution chain with just a clip-on lid and a paper sheet for protection and tamper evidence.

by a process called injection moulding. Injection moulding is able to form the very precise shape required that allows the push lid to be held in place by friction of the lid against the side wall of the container it is reclosing.

10.3.2 The dead fold system

Another simple system that has been around for years is one often used on 'bag in a box' packaging systems such as those used for breakfast cereals. The bag is sealed at the factory and the seal is opened at first use. Thereafter the cereal is kept as fresh as possible by folding the bag back down into the box to provide some level of barrier (Fig. 10.3). The dead fold system only works with packaging materials that don't want to spring back and undo the folding that is providing the protection to the product. A similar approach is used by people without airtight containers to store biscuits after the packs are opened. The end of the pack is twisted to form a dead fold seal to slow down the migration of moisture into the pack. This is acceptable for short periods – and let's face it, an open pack of biscuits usually gets eaten pretty quickly anyway! The final consumer method when it comes to closing packs is often used on sliced bread packages. The bread is originally sealed using an adhesive tape around the neck of the polyethylene bag. Once the tape has been removed the bag would have a tendency to open, allowing moisture to leave the bag, and the top slices of bread would dry out.

Fig. 10.3 The dead fold relies on packaging materials that do not want to spring back and unfold themselves but is an inexpensive way of closing a pack once opened. Here is a 'bag in a box' packaging system for icing sugar. The paper liner has a polyethylene layer to reduce moisture migration but it's the paper layer that gives the liner its dead fold ability.

The simple method is to fold the polyethylene over and use the weight of the bread to prevent the fold from opening.

10.3.3 The sticker system

This is an enhancement of the dead fold system that is used for packaging materials that have poor dead fold characteristics. Seen on some large snack food packs and sometimes on premium bread products, this system provides an adhesive, peelable, sticker to hold the fold in place once the original seal on the pack has been opened. The application of a sticker is a technique to provide a reclosable pack without having to compromise on the excellent barrier and other properties of the packaging material. Another variation of the sticker system is the neck clip. Often seen on bakery goods, the neck of a preformed bag is held closed by a plastic clip, or sometimes a plastic strip with metal wire to add stiffness and the ability to hold a deformed shape. The sticker system has also been extended recently by packaging companies that have created an adhesive that can be peeled and restuck up to 20 times. The sticker is used to cover a hole in the packaging and can be peeled back for initial access and then simply reclosed as required (Fig. 10.4). This system does not, however, return the packaging to its original state. Once opened, the shelf life of the product will be affected.

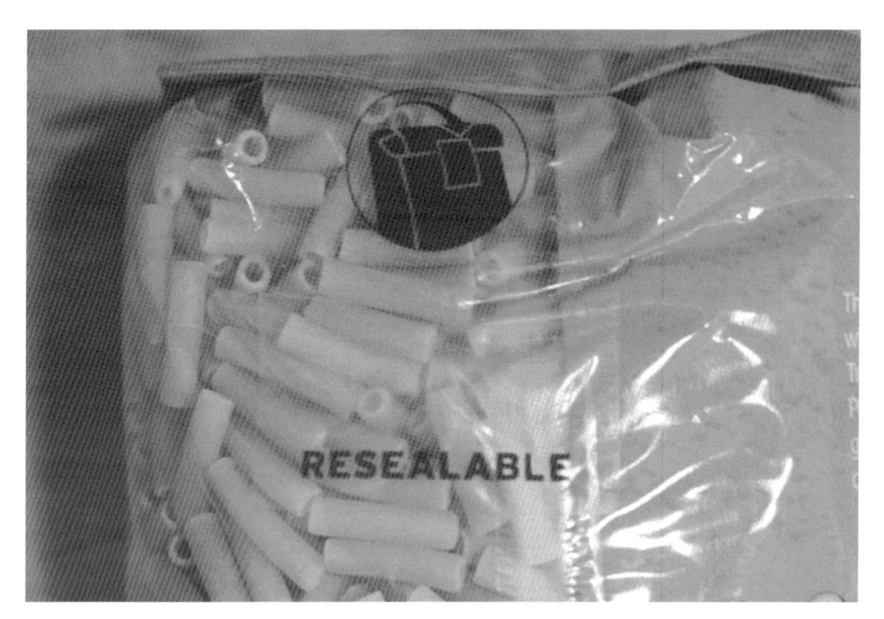

Fig. 10.4 The sticker system to 'add' resealability to a pack of pasta.

10.3.4 The zip system

This is the first of a group of systems where a specific device has been incorporated into a package to add reclosing as a feature. A good example here is the zip devices used on block cheese and grated cheese packs. A zip device is an expensive addition to a pack and so this type of feature is usually only used on expensive products that can pay for the feature (Fig. 10.5).

10.3.5 The velcro type system

Recent innovations have been made to incorporate a simple-to-operate hook and loop type system to reclose a pack. This system is more reliable than some of the zip systems, which are susceptible to damage and contamination that would occasionally make them difficult, or impossible, to operate. The hook and loop systems use well-tried technology to reclose the pack and are used on packs of cheese where major leaks in the pack (which can be caused by zip type systems popping open) would cause the cheese block to dry out.

There is a group of reclosure systems for bottles and these have developed over the years to become easier to use.

10.3.6 The screw cap

The traditional place to find screw caps was on everything from ketchup bottles to coffee jars and from fizzy pop bottles to toothpaste tubes. But the screw cap has been a revelation for the wine industry after years of trying to cope with the

Fig. 10.5 Here is an example of a zip closure. They do work but need a bit of care from the consumer to ensure that they are correctly closed. There are several different reclosure designs based on this principle.

technical issues caused by their cork and foil system of sealing. The industry's perception that customers would think less of the wine if it came with a screw cap has been proven to be completely wrong for the vast majority of consumers. This 'innovation' has occurred alongside a massive increase in wine consumption and it is one illustration of the power of easy opening and resealability on a product (Fig. 10.6).

Screw caps have also expanded their usage into areas such as fruit juice and milk cartons, bringing ease of opening and resealability (Fig. 10.7). The attraction of multi-use cartons of liquid product has seen a big increase in their use as brands fight for market share and shelf space.

Screw caps have also appeared recently on so-called 'fridge packs' of baked beans. The humble can is gradually being replaced by the multi-use package – the 'open it and use it all' systems that have been available for so many years are making way for more convenient packaging systems.

10.3.7 The flip lid

The screw cap has in some areas been replaced by the 'flip lid' where the resealability has been designed into an easy-to-open and easy-to-close feature (Fig. 10.8). Flip lids have appeared on ketchup bottles and table salt containers as well as in a novel device fitted to UHT milk and juice cartons (Fig. 10.10). The manufacturer's seal is required to prevent tampering with the product in the supply chain, but once that seal has been broken, it is the flip lid that recloses the container.

Fig. 10.6 The introduction of screw caps to wine bottles has coincided with a large increase in wine consumption. The easy-to-close screw cap helps keep the wine from oxidizing and allows people to leave some for tomorrow (well some people anyway!).

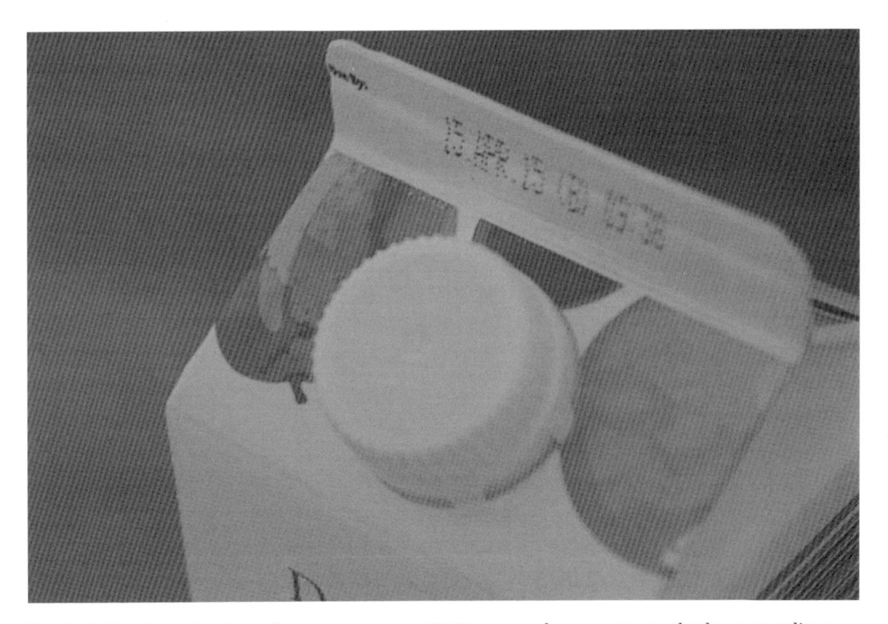

Fig. 10.7 The introduction of screw caps onto UHT cartons has overcome the long-standing problems associated with both gaining access to such cartons and reclosing them after first use.

Fig. 10.8 The flip lid is seen as an easy-to-use system for consumers to reclose a multi-use product such as ketchup. The flip lid is not the primary seal in these systems. That is usually provided by a foil or other material that is sealed to the container under the flip lid and is removed at first use of the product.

10.3.8 Sports cap

This is an extension of resealable systems that is designed to meet a well-defined problem. Standard bottle screw caps take some effort and time to undo and reclose if people want to take a small drink from a bottle. Once the factory seal is broken, sports caps are quick to open and close and they immediately make the bottle spill proof when in the closed position (Fig. 10.9). The sports cap has enabled soft drinks companies to change the way in which their customers interact with the product and this has allowed the growth in appeal of sports nutrition drinks in a very crowded marketplace.

10.3.9 Resealing using adhesives

This is a useful technique that can be applied to flexible packaging materials. It relies on the interaction between a pressure sensitive adhesive and cold seal adhesive. A thin layer of pressure sensitive adhesive is positioned around the required seal area. Next this layer is covered with a layer of cold seal adhesive in exactly the same position so that all of the pressure sensitive adhesive is covered. A layer of cold seal adhesive is then applied to the other packaging material that is to be used as the 'lid'. The adhesives are carefully selected so that the bond between the pressure sensitive adhesive and the substrate is the weakest of all the bonds. When the pack is opened the pressure sensitive adhesive remains on the 'lid' of the packaging but because of its pressure sensitivity it can easily be used to reseal

Fig. 10.9 The reclosable sports cap is designed to be very easy to use and so is very good for people who need to open, use and then close a package rapidly. The sports cap is a costly addition to a package and so its use tends to be on products that can cover the additional costs.

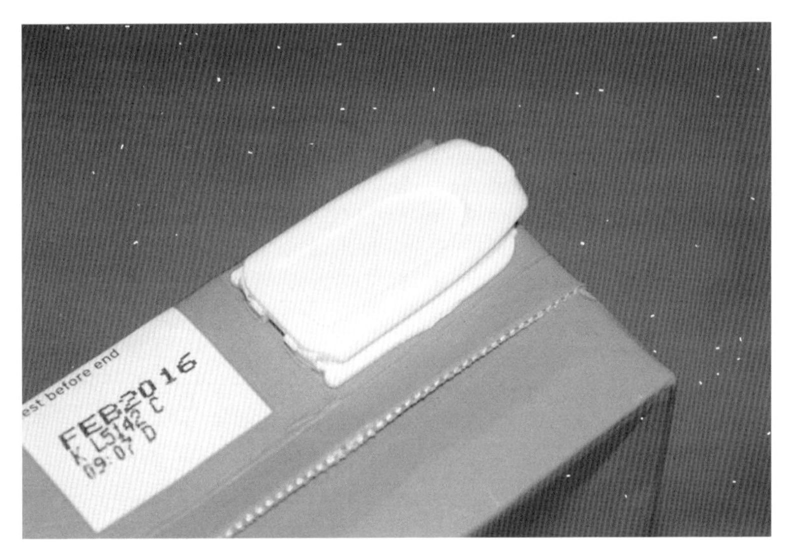

Fig. 10.10 The use of expensive injection moulded lids for rigid containers allows the product to incorporate more benefits than simply being reclosable after initial opening. Here the system aids pouring and minimises spills.

the pack a number of times. The peel and reseal systems have effectively created sticky labels that are able to cover over an opening in a package. More and more products are starting to use this as a way of adding reclose features to packs that didn't have them before.

Fig. 10.11 Single portion packs are an alternative to reclosable packs but use larger quantities of packaging materials and therefore are more expensive to produce and more costly to the environment.

10.3.10 Injection moulding

The final group of resealing systems are those that are commonly associated with pre-packed spices. These expensive items can be packaged this way because the cost of the pack can be absorbed in the overall cost of the high-value product. The resealing feature here is also designed to help with the controlled dispensing of the product, so the closure serves a dual purpose. An injection moulded twist system is used to open the pack, commonly a small glass jar, sufficiently to allow small quantities of the contents to be dispensed. Injection moulded parts on a package can add a whole set of benefits to the product and there are many examples where this technique has been used.

10.4 Single portion packs

With almost all resealing systems there is a need for a stronger, and better, seal on the package to provide the barrier properties and tamper evidence required for the product, but resealable multi-use packs are seen as a viable alternative to single portion packs, which, in a lot of cases, use an unnecessary quantity of packaging materials when a resealable multi-use pack would be a better option from an environmental point of view. However, single portion packs do have a place in the market, and food companies are able to use them to increase the appeal of their products (Fig. 10.11).

CHAPTER 11
Seals for non-flexible packs

11.1 Introduction

This chapter will look at the seals that are required and the methods used to achieve them in non-flexible packs such as glass bottles, jars and cans. Non-flexible packaging made of glass and metal is often associated with long shelf life products that have been subjected to some kind of post-pack processing or products that are 'self-preserving' due to their pH or sugar content. For example, a can of food product will be subjected to a heat process after sealing in order to render the contents of the pack commercially sterile and capable of an ambient shelf life of up to 2 years. The post-pack processing involves the container being heated in a pressurised steam retort at temperatures of around 121°C. Because of the high temperatures needed for the process to be carried out on a reasonable timescale the container has to be held in a pressurised retort to stop it from exploding. So a can, and its seals, have to withstand all of the normal issues of a supply chain but also have to cope with high temperatures and pressure differentials during the filling, sealing, heating and cooling processes.

11.2 Sealing systems for heat-processed cans

The ends on a standard can are held in place by a double seam (Fig. 11.1). This gives the can a lot of strength but does not, in itself, give a good seal. The seal of the can is achieved with the use of a flexible mastic layer that is able to migrate, as the double seam is made, to fill any voids and gaps in the seal. Without the addition of the mastic layer high-speed canning operations would not be possible and a large number of leaking cans would be made. Leaking seals on a can are a

Handbook of Seal Integrity in the Food Industry, First Edition. Michael Dudbridge.

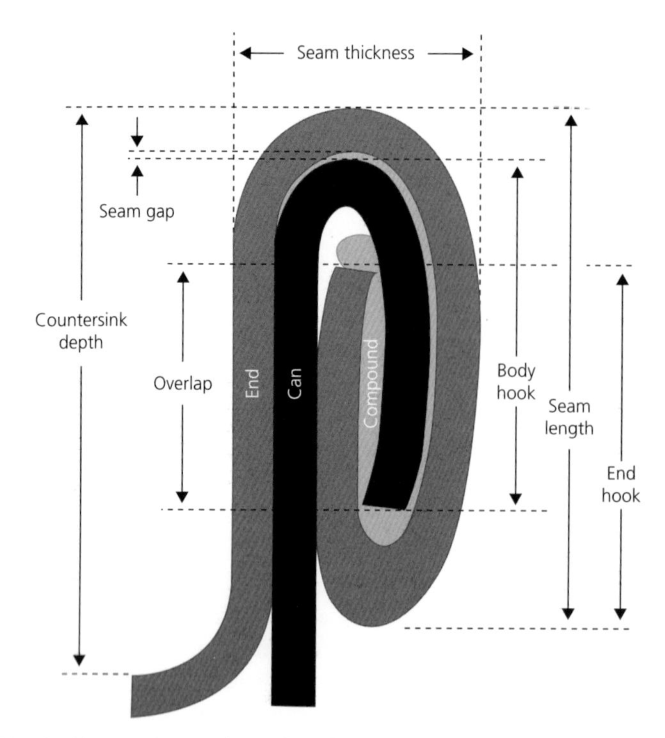

Fig. 11.1 The double seam is a mechanical method of fixing the end of the can onto the body. It does not make a seal without the inclusion of a mastic material (or compound) to fill the small channels and produce a hermetic seal.

particular issue as the presence of the seal is often the only thing that is stopping the sterile food from becoming contaminated with bacteria. For example, if the seal is leaking during the cooling process that follows steam retort sterilisation, contaminated cooling water can be drawn into the can as it cools. The bacteria introduced into the can in this way will have a potentially perfect environment in which to grow and will either spoil the food, or worse, cause food poisoning of the consumer.

One cause of a leaking seam on a canned food product that becomes apparent during its shelf life is the growth of micro-organisms inside the can. This causes the can to become pressurised and results in blown cans. The microbes are present either through the can being underprocessed or because the can leaked during the cooling cycle, drawing in cooling water and contamination. During the cooling process the metal of the can contracts and this reseals the can, trapping the micro-organisms inside the can. As the micro-organisms grow they produce gas, which causes the expansion rings on the ends of the can to move, giving a blown appearance.

11.3 Glass jar sealing systems

Glass jars can also be used with post-pack heating processes to obtain an extended life at ambient temperatures, but a large number of glass jars are 'hot filled' and the lid is fixed onto the jars while the temperature inside is still above 80°C. The hot filling process is often used for products such as jams and chutneys to produce a product that is free from pathogenic organisms. Non-pathogens that might survive the heating of the product are unable to grow because of the conditions within the food. For example, a jar of jam is hot filled and organisms are unable to grow because of the high sugar content of the foodstuff. Other products may be protected by high levels of acidity in the food or drink or a combination of different preservation techniques that together give the food a long shelf life at ambient temperatures. This is known as 'hurdle technology' but one vital component is the packaging and the quality of the seal. The lid is held in place with a feature that locks it against other features moulded into the neck of the jar. The lid is usually locked with a quarter turn that pulls it onto the jar to deform a mastic layer placed inside the lid (Fig. 11.2). By compressing the mastic layer a seal is made while the contents of the jar are still hot. As the jar cools, a partial vacuum is formed in the head space inside the jar. Some jar lids have used this fact to build in a seal integrity test for consumers. An area at the

Fig. 11.2 The lid on a glass jar is held in place by a mechanical quarter turn which also compresses a mastic material to produce a hermetic seal. Here we can see the jar lid in place and the mastic material that ensures a hermetic seal when the lid is first closed and held in place with the mechanical fixing in combination with a partial vacuum inside the jar.

centre of the lid is slightly raised and this raised area, or button, is pulled down by the effect of the partial vacuum. If the seal is not good or the jar has been previously opened then the safety button pops up so that the consumer can tell if there is a problem with the product. Using the safety button to test the jar before opening is a useful feature and gives consumers confidence in the pack. Because of the partial vacuum inside the jar, any leaks of the seal will result in air, and potentially bacteria, getting drawn into the jar. Only when the pressure inside and outside the jar have equilibrated will the contents of the jar have an opportunity to leak out.

11.4 Bottle sealing systems

There has been an evolution in bottle sealing techniques over many years. One of the principal demands on a bottle seal is that often the contents of the bottle are pressurised and so the seal needs to be able to withstand the pressure differential that occurs (Fig. 11.3). Beer, cider and carbonated drinks are only possible in bottles (both rigid and semi-rigid) because the closure is capable of withstanding the internal pressure.

The original system was the wired cork system that is still typical for champagne and sparkling wine products. This is a simple adaptation to hold the seal made by the cork in place with stiff wire. This is an expensive option that is acceptable for expensive products but not for less expensive carbonated drinks, which would not be able to support the additional cost at retail. Different solutions were needed.

The crown cork was the first development of the champagne cork system. A thin layer of cork was held in place by a metal cap that was bent so it gripped the glass bottle. This was a good innovation still seen in the beer market but not as convenient for the consumers as some later innovations. To remove a crown cork a special tool is needed – the bottle opener is now common in most parts of the world.

The twist crown cap was developed to provide a seal that could withstand the internal pressure but be removed simply by twisting, thereby obviating the need for a special opening tool. The twist crown cap is applied in the same way as a standard crown cap, using the same machinery in the bottling plant, but can be removed without the use of a bottle opener. The seal is made using a mastic material rather than cork, and the move away from natural materials, with their inherent variation, added greatly to the performance of the system.

The above three systems of bottle closure have no method of reclosing the bottle after opening, so the contents of the bottle have to be used in one go. There was clearly a need to be able to reclose bottles, especially on larger sizes, and so a new development requirement was formed.

The screw cap was developed to address the requirements for easy opening and also resealing of a bottle. There are many variations in design of the screw

Fig. 11.3 Here are some bottle closures that reflect the wide variety of sealing systems that have been developed for this important packaging format. The variation depends on the presence or absence of internal pressure (from carbonated products) and on the requirement for ease of opening and reclosing.

cap but they all have a right-hand thread and are able to seal down onto the hard glass neck of the bottle. It is usual for there to be three turns between open and closed but there is a wide variation as the pitch of the screw is important in the ability of a screw cap to hold tight enough to prevent the escape of the contents of the bottle, both liquid and gaseous, and at the same time be easy to open for the end user.

A simplified screw cap is often used on glass bottles that have no internal pressure. Here just a quarter turn is needed to remove and refit the cap. This kind of design is often used in the packaging of ketchup in glass bottles.

11.5 Sealing methods for rigid containers

The machinery involved in the sealing of rigid containers varies greatly depending on the packaging format but can be broken down into basic groups.

11.5.1 Preformed container and preformed lid

This is the kind of system used in high-speed bottling and jar filling lines where glass (and increasingly rigid plastic) bottles and jars are produced to tight dimensional specifications so that they can be handled and positioned accurately as they pass through the machine. The bottles and jars enter the machine via the in-feed and are then filled while moving. The filled container then passes to the capping/lidding section of the machine where a preformed lid (also manufactured to a tight specification and typically made of a rigid plastic material or metal) is introduced and marshalled into position. The lid is placed onto the container and then tightened using a screw action. The design of the lid often contains a deformable mastic material that creates a hermetic seal.

11.5.2 Preformed container and lid moulded into position

This kind of sealing system is typified in canned foods. A preformed can body is delivered to the filling and sealing system. It is vital that the can body is very dimensionally accurate as this operation takes place at speeds up to 2000 cans per minute. The can is filled and then a lid is placed onto the can. A mechanical seal is then made using rollers to mould the lid (usually called the can end) onto the body of the can. This moulding operation is normally carried out in two stages: first the lid is hooked under the lip of the can body and then the seal is compressed together to ensure that the mastic material that is part of the lid design is squeezed and forms a hermetic seal. This kind of system can also be seen in the sealing of bottles where a screw cap is 'rolled' into position and formed onto the bottle as part of the sealing process. The lids, typical of

Fig. 11.4 Polypropylene injection moulded lid and base. The partial vacuum formed inside the pack as the soup cools helps keep the lid in position as well as the mechanical clip designed into the lid and base. Tamper evidence is also designed in, as the only way the lid can be removed is by breaking a small tab on the rim of the container.

the screw cap wine industry, are delivered partly formed, placed on the top of the bottle and then subjected to pressure to finally fit them to the neck of the bottle.

11.5.3 Preformed container and a clip-on lid

This kind of system is sometimes less reliable in terms of seal integrity but is typified by two groups of products. The first is coleslaw and yellow spreading fats. In this group the container dimensions have a slightly looser tolerance and there is some degree of stretch designed into the lidding material to compensate for the variation. This system keeps the packaging costs low as it means that the containers and lids can be manufactured using thermoforming, which is a lower-cost operation. Typically, the container in these systems is manufactured using polystyrene and the lid is polypropylene, but other combinations are used, depending on the requirements.

The second group in this category is typified by the containers used for chilled soups. The preformed containers and lids are manufactured to higher tolerances by a process called injection moulding (Fig. 11.4). The containers and lids are typically made from polypropylene, which is able to retain its precise shape even when heated. This is useful as chilled soups are often packaged at around 90°C to help extend the shelf life and to draw a partial vacuum inside the containers as cooling occurs. Packaging speeds for this kind of system are much slower than for canning and bottling operations.

CHAPTER 12
Oxygen and moisture migration

12.1 Introduction

In the early chapters of this book there was a discussion regarding what makes a good seal. The definition of a good seal is one that is adequate to retain the product and protect it from the environment outside the pack for a sufficient length of time that the quality and safety of the product is not affected.

At the limit of this discussion is the ability of moisture and oxygen to migrate through the packaging materials themselves rather than through an inadequate seal. The rate of movement of oxygen through a packaging material is called the oxygen transmission rate (OTR). The rate of movement of moisture through a packaging material is called the moisture transmission rate (MTR). This discussion is especially important when the safety or quality of the product can be adversely affected, and consideration of this issue may be necessary if unexpected consequences are to be avoided during changes in packaging materials. For example, reducing the thickness of packaging materials could increase the OTR or MTR and this would have an impact on the quality of the product at the end of its shelf life. A change of supplier of a packaging material could also impact the OTR and MTR.

With some products it is desirable to have a controlled movement of oxygen or moisture through a pack and this can help improve quality or extend shelf life of the product.

This chapter will look at all aspects of migration through packaging materials, including migration of oxygen, moisture, fats and other materials.

Handbook of Seal Integrity in the Food Industry, First Edition. Michael Dudbridge.

12.2 How does oxygen migrate through materials and how do we measure OTR?

Oxygen can move through materials by a process of diffusion at a molecular level. For any diffusion to occur there must be a concentration gradient, so oxygen will move into a pack if the atmosphere inside the pack is lower in oxygen concentration than that outside the pack.

Normal atmospheric air contains around 20% oxygen, so the oxygen concentration inside the package would have to be lower than that to cause any diffusion. The lower concentration inside the pack can occur in several ways. The pack could have been created with a modified atmosphere to deliberately reduce the oxygen content to a low level to protect the product from flavour changes, as is the case with snack foods, or to prevent the growth of mould (which requires oxygen to grow), as is the case with long-life bread products such as tortilla wraps.

Where a concentration gradient occurs oxygen can diffuse through materials at a known rate, which is a characteristic of the material and its thickness. Some materials provide a very high barrier to migration, even at very low thicknesses. Other materials provide very little barrier to diffusion and therefore should not be used when oxygen migration is an issue for the product.

PET, polyethylene terephthalate; PE, low density polyethylene; PVDC, polyvinylidene chloride; PVAL, polyvinyl alcohol; EVOH, ethylene vinyl alcohol are important materials with respect to OTR (Table 12.1).

The oxygen transmission rate (OTR) is measured in cm^3 of oxygen per m^2 of material per day when the temperature is 23°C and the relative humidity is 50% and there is a one atmosphere pressure differential across the film ($cm^3\ m^{-2}\ d^{-1}\ atm^{-1}$ at 23°C and 50% RH) and the moisture transmission rate (MTR) is measured in grams of water per m^2 per day when the temperature is 23°C and the relative humidity is 75% ($g\ m^{-2}\ d^{-1}$ at 23°C and 75% RH).

12.3 Packaging specifications

It can be seen that oxygen and moisture migration rates can vary greatly between materials because of their chemical make-up, their structures and even their manufacturing methods. It is vital for companies where OTR and MTR are important that these factors become part of the overall material specification. Without this there is a risk that, over time, the material may vary and as a result issues may arise with the final product. Measurements of OTR and MTR should be taken periodically to protect your business from potential problems. Changes in supplier or material specifications should always include considerations of MTR and OTR. Table 12.1 shows that with the addition of even quite thin layers of some materials the OTR and MTR can change greatly and this knowledge is used in the creation of multilayer laminated materials designed to meet desired OTR and MTR requirements along with other needs.

Table 12.1 Table of packaging materials and their resistance to oxygen and moisture migration.

Packaging material	Layer thickness (micron)	OTR	MTR
PET	12	110	15
PET / PE	12 / 50	1.24	0.37
PET / PVDC / PE	12 / 4 / 50	0.33	0.13
PET / PVAL / PE	12 / 3 / 50	0.13	0.39
PET / EVOH / PE	12 / 5 / 50	0.06	0.27
PET / metallised aluminium / PE	12 / - / 50	0.12	0.03
PET / silicone oxide	12 / -	0.06	0.06
PET / aluminium foil / PE	12 / 9 / 50	0	0

Source: reproduced from Lange and Wyser (2003) *Packaging Technology and Science* 16(4): 149–158.

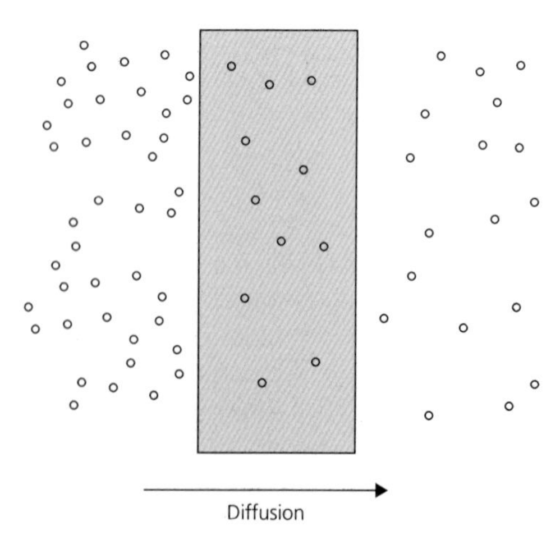

Fig. 12.1 A mechanism of gas or vapour transmission through a plastic film. The mechanism is made up of three stages. First, the molecules are absorbed on the surface of the packaging, then the molecules diffuse through the packaging material and finally the molecules evaporate from the surface. Interruption or resistance to any of these processes will increase the barrier properties of the material.

12.4 The theory of diffusion

In materials with no defects such as pinholes or cracks, the mechanism for gas and water vapour transmission through a film or coating is diffusion. This means that the oxygen or moisture dissolves in the film at the higher concentration side, diffuses through the film, driven by a concentration gradient, and evaporates from the other surface (Fig. 12.1).

The diffusion rate depends on the size, shape and polarity of the penetrating molecule and on the crystallinity, degree of cross-linking and polymer chain motion of the packaging material. These factors are also related to temperature, so the diffusion rate will vary depending on the storage temperature of the package. Diffusion into semi-crystalline polymers is confined to the amorphous regions because the crystalline structures are fixed with very few paths for the diffusing molecules. This helps explain the relative differences in OTR and MTR for thermoplastics of different structures. Highly crystalline materials such as ethylene vinyl alcohol and silicone oxide are almost perfect barriers even at very low thickness, whereas less crystalline materials such as amorphous polyethylene terephthalate can be poor barriers even when the material is quite thick.

12.5 Measurement of OTR

The OTR of a flat piece of packaging film is measured on an instrument that is operated in a very controlled way to ensure consistent results. A sample of the film is first inspected for faults and it is then clamped into a machine that allows the introduction of a high concentration of oxygen to a known surface area on the top side of the film. The oxygen concentration in a chamber below the film is measured and because the chamber is of a known volume the quantity of oxygen passing through the film can be measured. In order to get consistent results the testing is carried out in a room with a known temperature and relative humidity as both these factors can impact the transmission rate (Fig. 12.2).

As you would expect, ASTM International (formerly the American Society for Testing and Materials) has a standard for testing materials for oxygen transmission. Its latest standard to lay out the testing methods for determining this very important measure of packaging performance is ASTM 2622-08 (2013).

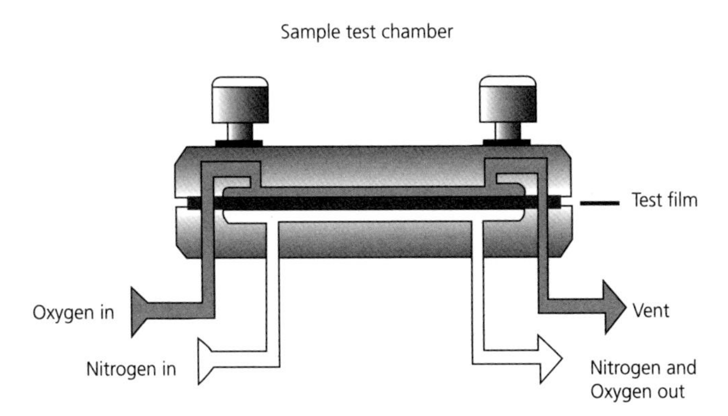

Fig. 12.2 OTR measurement instrument where the OTR is measured in a controlled environment so that comparisons can be made and decisions taken about the barrier property.

There are several instruments available that have been developed to measure the OTR using different types of sensor but they all have one issue, which is that they test a flat piece of packaging film. There are also a few systems that allow the testing of oxygen migration of a full package including seals and shaped trays.

12.6 MTR

The MTR of a packaging material, sometimes called water vapour transmission rate (WVTR), can be vital to the quality of a product at the end of its shelf life. Sometimes the moisture needs to be kept out of the package and sometimes it needs to be retained inside the package to ensure that the product quality remains high.

MTR is a very important packaging characteristic for the food industry with a wide variety of products requiring protection.

Dry products, such as biscuits, dried pasta, dried soup mixes, instant coffee, snack foods and breakfast cereals, need protection from moisture ingress into the packs during long ambient shelf life. If the MTR of the packaging material is too high then these products, with their low levels of water content, will attract moisture from the environment and as a result product quality will deteriorate.

Products with high water content, such as some fresh vegetables, ready meals, cakes and bread products, need the moisture to be retained inside the pack to prevent the products from drying out.

There is a special case for chilled and frozen foods. In chilled and frozen conditions the relative humidity of the air around the products is very low. So the moisture gradient between the food product and the air around it is very large and rapid dehydration can occur if the MTR of the packaging material is too high. Chilled products tend to have a relatively short shelf life, so the effect is not always significant, but frozen products can have a 'use by' date of up to 2 years, so dehydration can be a major issue at the end of shelf life. The MTR of materials used in the frozen food industry has to be low to prevent a quality fault known as freezer burn, where the dehydration is so large that the food product takes on the appearance and texture of a product that has been overcooked (Fig. 12.3). At freezer temperatures water molecules are less mobile than at 23°C but there is an amount of 'free' water not frozen into ice crystals and this is the water that is able to leave the food if the barrier properties of the packaging are inadequate.

The issue of MTR is significant for a large range of food companies and so needs to be considered very carefully when changes are being made to packaging. It also needs to be considered for products that are 'suitable for home freezing', where the packaging has to be able to cope with the demands of chilled environments and also frozen storage conditions.

The mechanism of moisture transmission is very similar to that of oxygen. A moisture gradient will encourage water molecules to diffuse through the packaging material. The rates of migration can be seen for various material combinations

Fig. 12.3 A frozen beef burger that is showing signs of freezer burn after being packaged in a material with a high MTR for an extended period.

in Table 12.1, where, once again, it can be seen that different materials have different barrier properties with respect to moisture migration. What is the MTR for your packaging materials? Is MTR important for your business?

Other materials than water and oxygen are also able to diffuse through packaging materials. One troublesome type in the food industry is the diffusion of fats and oils. This causes a great deal of problems in products that contain high levels of fat as it can interfere with label appearance and printing. Other materials that are very important for food manufacturers are the migration of odours and even printing inks. All of these need to be considered.

12.7 Impact on MAP systems

Modified atmosphere packaging (MAP), as the name suggests, uses the modification of the atmosphere inside a package to enhance the quality of the foods or extend the shelf life. It is vital that the packaging materials used in a MAP system are able to retain the modified gases inside the pack for the entire shelf life of the product. If the atmosphere is lost or diluted in any way then the protection is lost and the food could deteriorate or even become unsafe to eat. Good seal integrity is vital in MAP systems but so too is the performance of the packaging material in terms of its barrier to gas diffusion. Typically, a MAP pack contains the same gases that are usually found in the normal atmosphere but the proportions are changed to bring about a desired effect.

The atmosphere used will depend on the effect required and the product being packaged but typically will be an elevated level of carbon dioxide and a reduced level of oxygen. The barrier to carbon dioxide and oxygen migration is therefore very important if the atmosphere is to be preserved. For example, snack foods are typically packaged in an atmosphere that is very low in oxygen to prevent flavour changes in the product, so a packaging material that has a very high barrier to oxygen (i.e. a low OTR) is used. This in recent years has been a complex laminate packaging material that contains a metallised layer to provide a high barrier to oxygen and prevent it migrating into the package. The metallised layer is very thin but has the required barrier and also acts to prevent ultraviolet light from entering, which can cause rancidity of the frying oils contained in the products.

12.8 Controlled atmosphere packaging

Controlled atmosphere packaging (CAP) controls the migration of gases and moisture through careful selection of packaging materials and in some cases the micro-perforation of the packaging material. In a book that is focused on seal integrity and the creation of sealed packages it may seem strange to be referring to packages that are designed to leak, but this is what CAP is.

Certain products, such as vegetables, require access to a controlled quantity of oxygen so that they can continue to respire and keep their cells in good condition. Too much oxygen will allow the respiration to occur too quickly, converting the starches in the vegetables to sugars and using up the water inside the product. Too little oxygen and the respiration will become anaerobic, leading to cell breakdown and poor quality product. The level of oxygen available needs to be controlled to keep the vegetables in the best possible condition for as long as possible.

The level of oxygen migration can be very precisely controlled by the careful selection of the correct packaging materials and, sometimes, the use of micro-perforation of the packaging materials to provide a faster route for oxygen migration. For example, the packaging of prepared and washed salad leaves could not occur without a full understanding of CAP. Too little oxygen reaching the leaves and they would soon wilt as they used up the available oxygen and then continued to respire anaerobically. Too much oxygen and the respiration rate would be too high, meaning that the leaves would soon exhaust the water available and start to turn brown in colour. The whole of the business of supplying washed and pre-prepared leafy salads can only occur with a shelf life that allows the supply chain to operate at reasonable waste levels. Without CAP the supply chain would generate too much waste and as a result it would become uneconomic.

12.9 Impact of down-gauging of material thicknesses

The environmental impact of food packaging is well documented. The size of the industry means that even small changes in the packaging weight used can have a great impact. For this reason there has been a shift in thinking in the food packaging sector towards lighter weight food packages. The lighter weight is often achieved by making the packaging materials thinner. This certainly reduces the weight of each package but there are other impacts that should also be considered.

12.9.1 Increased MTR and OTR

This is an obvious consideration but the impact of reducing packaging thickness can have unintended consequences if the impact on OTR and MTR are not fully assessed. There are many examples where the savings achieved through a thickness reduction have been outweighed by an increase in product quality issues towards the end of shelf life.

12.9.2 Packing machine operation

Reducing the thickness of packing films can have an adverse effect on the way that the film handles on the packing machine. This can result in higher waste and reduced efficiency so that the savings, in cost and environmental terms, are soon eroded.

12.10 Barrier layers to improve OTR and MTR

A long time ago it was a feature of frozen food products that the packaging was made up of waxed cardboard. A similar technique was also used for white sliced bread when it first came onto the market. The wax layer on the packaging materials was used to reduce moisture loss and protect the frozen foods from freezer burn and prevent the bread from drying out. The principle of using thin layers of a material to add extra capability to a package has been with us in other areas too. Waxing the outside of a cheese to prevent mould growth is another example.

Modern packaging technologists have many more opportunities to change the performance of a packaging material by the addition of thin layers into a multilayer system.

Table 12.1 at the start of this chapter highlights some of the materials commonly used for their barrier properties. Each of these materials is expensive, so the thickness of the layers used is often minimised to keep the overall cost of the material under control.

Materials such as polyvinylidene chloride (PVDC), polyvinyl alcohol (PVAL) and ethylene vinyl alcohol (EVOH) are commonly used to bring extra performance in terms of barrier properties to other systems where barriers need to be boosted.

CHAPTER 13
Tamper evidence technology

13.1 Introduction

With all kinds of reclosable packaging one of the main concerns of consumers and end users is the integrity of the package just before it is opened. Many types of tamper evidence technology have been developed to try to ensure that if a pack has been opened it is very obvious to the consumer so that appropriate action can be taken. Tamper evidence technology on glass bottles was among the first to be developed to help protect consumers. When a pack is not reclosable the primary packaging provides all of the tamper evidence required. The issue here is with packages that are designed to be reclosed during their use.

13.2 Neck sleeves and paper bands

These are typical on corked wine bottles. The cork is the main source of a seal but the sleeve is made of a material that is able to improve the barrier to the movement of oxygen into the bottle (Fig. 13.1). Traditionally, foil was used combined with a paper sealing band, but this was replaced by heat-shrunk thermoplastic materials. Some bottles use simple paper labels that are fixed to both the bottle and the cap, so that when the cap is removed the paper label is damaged and it becomes obvious that the bottle has been opened.

Here are a couple of examples of label and sleeve systems that provide tamper evidence.

Handbook of Seal Integrity in the Food Industry, First Edition. Michael Dudbridge.

Fig. 13.1 A neck sleeve typically used to seal a bottle. The seal is not to provide any barrier properties but is there to assure the consumer of the product that the container has not been opened during distribution or retail sale.

13.2.1 Tamper bands

These are fixed around the neck of a bottle (and sometimes pots and jars too). The bands are made of a thermoplastic material that shrinks when heated, so after the cap is placed onto the bottle a tamper band is added and then the bottle is sent through a heated tunnel. The heat shrinks the band so that the cap cannot be removed without first breaking the tamper band (Fig. 13.2).

13.2.2 Tamper evident screw caps

These were developed to aid high-speed bottling machines by removing a step from the process. The cap is attached to a tamper evident section and both are fitted at the same time as the bottle is closed. On opening, the screw cap is twisted in the normal way but, because of its design, the tamper evident section cannot rotate. As a result the cap and the tamper evident section break apart to provide the evidence to the consumer that the bottle has been previously opened.

13.3 Induction sealed tamper evidence systems

This technique is used on large semi-rigid bottles of milk and also on ketchup bottles (Fig. 13.3). Under the screw cap on the top of a bottle is a membrane that has been sealed to the bottle using induction heating. To gain access to the

Fig. 13.2 A tamper band is used to provide a level of assurance to the consumer that a product has not been opened during distribution. In the case of a clip-on lid container, such as a coleslaw pot or a large pot of yoghurt, the tamper band also serves to prevent the clip-on lid from popping off during distribution. No additional barrier is provided, and indeed in some products such as coleslaw the tamper band allows gas to escape from the pack if the uncooked vegetables in the product start to ferment and produce carbon dioxide. In this type of coleslaw product a perfectly sealed pack may well inflate and split open if the gas pressure inside increases due to higher than normal storage temperatures.

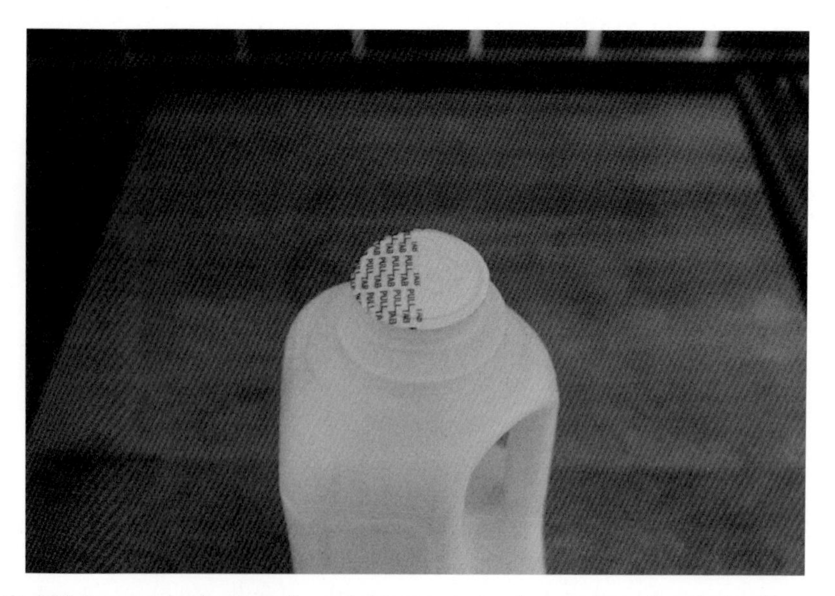

Fig. 13.3 An example of an induction sealed tamper device. Commonly used on plastic milk bottles and squeeze-type ketchup bottles these devices are used under a cap and they allow the product to become multi-use, with the screw cap or flip lid providing the closure when required.

product inside the bottle the outer cap is removed and then the membrane is peeled back. The outer cap is then used to reseal the bottle. A foil and thermoplastic membrane is incorporated into the packaging of jars of coffee to provide a high level of barrier to oxygen but also to provide the tamper evidence required by consumers.

13.4 Injection moulded parts

Polypropylene containers that are semi-rigid are often formed by a process of injection moulding. This gives a great opportunity to build tamper evidence into the design of the package (Fig. 13.4). Typically used for hot filled soup products, the lid is clipped into place and it is only possible to release the lid by breaking a small tab on the lip of the container. Once the tab is broken it is possible to remove the lid, but the broken tab remains broken to provide the clue to the consumer that the package has been previously opened. This technique is also used in the caps of plastic vegetable oil bottles where a pull strip is incorporated into the cap design to make the pack easy to open and also provide the required tamper evidence.

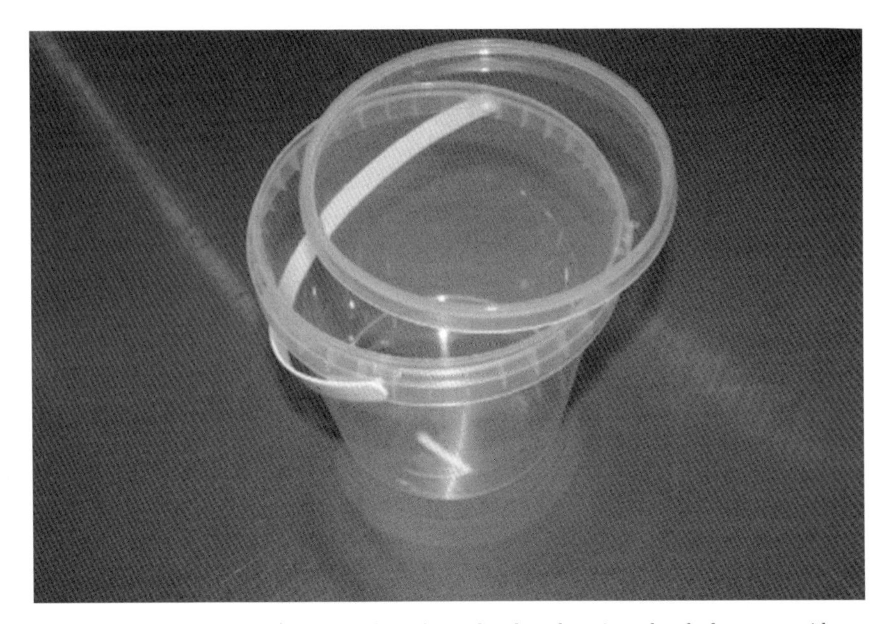

Fig. 13.4 There are lots of devices that have been developed to give a level of tamper evidence to semi-rigid packaging. This group of tamper evidence devices relies on the precise shape of the components and so they are often injection moulded. The price of these systems is quite high and so will generally be seen on products that can shoulder the extra expense.

Fig. 13.5 The volume of production of these injection moulded products helps keep their cost down. It is apparent that all manufacturers of fizzy pop bottles utilise the same size of bottle cap for their standard high-volume products. This is partly due to the cost and availability of the caps with their built-in tamper evidence band that breaks away from the main part of the cap when the bottle is opened.

13.5 The screw cap with band

To make the addition of tamper evidence devices easier for the packing company a technique was developed that allowed the tamper evident feature to be added to the package at the same time as the cap. Used on everything from a 2 l bottle of fizzy pop to motor oil, the cap is put onto the container in the standard way with a clockwise rotation. The cap locks into place and can then only be removed by breaking its attachment to a band. As the cap is removed the band stays in place, so it is immediately apparent to the consumer that the container has been previously opened (Fig. 13.5).

13.6 Test buttons on jars

Typically used on jars that have been filled hot, a 'button' is built into the metal lid. As the product cools after the container is sealed the button is pulled in by the partial vacuum formed inside the jar (Fig. 13.6). When the pack is opened, usually by a twisting action on the lid, the button pops up indicating that the seal on the

Fig. 13.6 When a product is filled hot into a rigid container and then the container is sealed there will be a partial vacuum present inside the container when it cools. This is due to the contraction of any air and the condensation of any steam that was in the container when it was sealed. This fact has been used to develop a tamper evidence device in the centre of the cap. A button shape is designed into the cap that is pulled in by the partial vacuum. If the button pops out it is an indication that the seal has been broken.

package has been broken. The button then becomes an easy test for the user to check that all is well with the product. The button is often designed so that if a pack has been previously opened the button makes a 'click' when pressed to indicate that it should be treated with caution.

13.7 Film overwrap for cartons

Typically used on boxes of tea bags and chocolates, a cardboard carton is overwrapped with a film layer that is sealed (Fig. 13.7). The only way into the pack is to remove the overwrap and it then becomes obvious that the pack has been opened.

13.8 Stickers and paper labels

These are seen as an inexpensive method of giving tamper evidence to a package. The pack is closed in the normal way and then the opening is sealed using a sticker or glued paper label (Fig. 13.8). The pack can't then be opened without the label

Fig. 13.7 The film overwrap is used to form a tamper evident layer on the outside of a package. The materials used for this layer will be selected to also impart some barrier properties and will not have a peelable type of seal. The only way into the pack is to destroy the layer, maybe with an easy-open rip strip.

Fig. 13.8 A paper label is often the preferred tamper evidence device for low-volume products where some of the more expensive options cannot be justified. Often used by small companies, paper labels are a convenient way of adding tamper evidence to a pack.

being removed or broken and so the end user can be reassured that the pack has not been tampered with during distribution or display.

In the next chapter there is some discussion about the use of colour-changing pigments to indicate if a pack has been opened. This is being developed for a new range of reclosable packaging systems aimed at the sliced cured meat sector, where modified atmospheres are used to extend the shelf life of the products and the resealable packages are convenient for consumers.

CHAPTER 14

Innovations in packaging to improve seal integrity and the detection of sealing issues

14.1 Introduction

Seal integrity is such an important issue within the packaging industry that there is a constant stream of innovative ideas either to improve the seal integrity of packages or to detect an inadequate seal once it is made. The main causes of seal integrity issues were discussed in Chapter 5, so here a few recent examples (in some cases yet to be perfected) of innovative thinking in the area of seal integrity. The examples are grouped into categories to make it easy to see the field being looked at, and this also neatly identifies which section of the packaging industry is working on each particular innovation. It can be seen each year that innovation in packaging is a major feature of the packaging industry. Massive exhibitions take place all over the world to showcase the latest innovations and these exhibitions are used by industry to ensure that optimal packaging performance is obtained. Often, of course, innovative products that are new to the market carry with them a risk that they will not work, and they can also add considerable expense to the product being packaged, but that is a balance that has to be judged by the factory when deciding if the new idea is right for them and the supply chain of which they are a part.

14.2 Packaging materials

In this section we will look at one particular innovation in packaging materials and packaging design that is trying to improve seal integrity and therefore the performance of the companies employing this innovation in their packaging systems.

Handbook of Seal Integrity in the Food Industry, First Edition. Michael Dudbridge.

14.2.1 The LINPAC Rfresh Elite® system

This is an example of recent innovation in packaging and material design to improve sealing performance. Traditional polyethylene terephthalate (PET) trays for use in the meat and poultry sector are constructed with a thin inner layer of polyethylene to act as a sealing layer. The Rfresh Elite tray is made of just PET, which improves its ability to be recycled. PET is a difficult polymer to seal at high speeds, so this issue is overcome by use of a sealing compound that is applied just to the sealing area of the tray. This sealing compound is designed to have a low softening point and as a result sealing times can be reduced. The compound can also be coloured in a way that makes it easier for vision systems to spot a faulty seal. This innovation has only been made possible by the use of a high-speed printing technique to apply the compound as the trays are manufactured.

14.2.2 Packaging shape and usability

There are lots of innovative shapes of package hitting the shelves of the retailers as they try to retain their market share in a very competitive situation. Packaging shape and usability are seen as being key drivers for shoppers, along with a reduction in the quantity of packaging material being used. Anything written here would be almost instantly out of date, so for innovation ideas for your packaging just take a walk around a retailer and look at what is appearing there. Better still, call your packaging supplier and get them to tell you what they think will be next year's big thing.

14.3 Sealing systems

In this section we will look at sealing machines and systems that are being developed to improve sealing performance with respect to seal integrity and predictable seal strength.

14.3.1 Laser sealing

The creation of a seal using energy derived from a laser system has been developed to try to bring about a step change in sealing performance for preformed trays. A conventional sealing system relies on the use of shaped and highly engineered components in a tray sealing machine. These tools are costly and have to be changed each time the shape or size of the tray changes. This adds downtime to the sealing operation as well as cost to the launch of a new tray shape.

Laser sealing does not rely on engineered components to change tray shape. The flexibility is built within the software controls of the laser, so change from one shape to another requires a software change that can be triggered using computer vision systems. A tray arriving at a laser sealing machine is scanned and the information is sent to the laser so that sealing can occur.

14.3.2 The integrity seal

This sealing system is designed to overcome several sealing issues and produce top and bottom seals on vertical form fill seal (VFFS) bags. The concept is that the sealing bars are replaced with a sealing wire, which is much narrower. The sealing wire is not heated all of the time, only instantaneously when a seal is being made. This reduces energy consumption and the issue of carbon build-up on the heated surface. The wire presses the packaging material layers into a deformable surface and seals them at the same time as cutting through them. This is achieved by generating temperatures that are above the melting point of the film. The system produces a very narrow weld seam that is not impacted by seal area contamination.

14.4 Detection systems

Here we will look at innovations in the way in which packs with inadequate seal integrity can be detected.

14.4.1 Seal scope

This is a system to detect product in seal during the sealing process. As two sealing jaws come together to form a seal on a VFFS machine there is a pattern of vibration and deceleration that can be recorded and analysed to produce a model of what good looks like. Any product that is trapped in the seal area as the seal is made will cause the deceleration and vibration to be different and this can be spotted by the computer controlling the seal scope system. Product in seal is the biggest cause of seal integrity issues and this system can alert the machine operators to a problem.

The seal scope system is also able to add another benefit to a production line.

VFFS machines (bag makers) are often combined with weighing and feeding systems that are linked electrically to the packing machine. On high-speed lines the bagging machine may be operating at speeds of up to 200 packs per minute and the coordination of the weighing and bagging machines is key to getting the product into the bag just before the top seal is made. If there is a slight delay in the product falling from the weighing machine the product will get trapped in the seal area, triggering the seal scope and the pack rejection system. The seal scope is not just a policeman rejecting faulty packs; it is able to look for patterns and feed information back to the weighing machine. So, for example, if the time it takes the product to fall from the weigher to the bag increases slightly and faulty seals start to be made, the seal scope can change the setting on the weigher to compensate, slightly increasing the time delay between weigher discharges to allow more time. On the other hand, if the seal scope is not detecting any issues with product being caught in the seal it can automatically shorten that delay and so speed up the packing process. So the seal scope is capable of optimising the packing rate of the system at the same time as protecting against product-in-seal faults.

14.4.2 Modified atmosphere packaging

Modified atmosphere packaging is a major part of chilled and other food supply chains. Without modified atmospheres some products would lose shelf life and the product would cease to be commercially viable. Seal integrity is vital in products where a modified atmosphere is used to help preserve the product, and so it will be no surprise that methods are constantly being developed to spot leaking packs in this area. Recent innovations have resulted in the introduction of a tracer gas into modified atmosphere packs purely to be used to detect leaks if they occur. The gas introduced needs to be in a low percentage and also to be easily detected in the event of a leak. Ideally, the tracer gas needs to be rare enough that it does not occur in normal atmospheres, so that if detected it can only have originated from inside the pack.

Hydrogen is a gas that has been used in tracer gas leak detection. The pack needs to contain around 3% hydrogen and this can then be reliably detected down to around 10 parts per million in the atmosphere around the pack to indicate a leak.

Carbon dioxide has also been used to detect leaks. It has the advantage of already being a part of many modified atmospheres, but it has a disadvantage in that a small quantity of the gas is always present in the air and so false positives can be a problem.

14.4.3 Carbon dioxide-sensitive labels

Recent developments in pigments and printing inks have opened up the possibility of creating a colour change in a pigment when it is in the presence of carbon dioxide. So a correctly packaged modified atmosphere package will have a label of a certain colour. If the package is leaking and the modified atmosphere escapes, the colour will change to indicate that the pack is leaking.

14.4.4 Multispectral imaging

The use of high-resolution images coupled with computer image analysis has been developed to allow the seal area of a package to be examined automatically and faulty seals to be rejected. Visual inspection of a seal often reveals anomalies in the seal area, but human vision can only carry out this inspection within the visible spectrum. Multispectral imaging can capture images for analysis beyond the visible spectrum in both directions, look for anomalies and provide information to packing machine operators (Fig. 14.1).

In the infrared spectrum it is possible to look for hot spots and cold spots in the seal area soon after the seal is made. A hot spot or a cold spot would indicate that the seal was not normal and that corrective action may be needed. The use of thermal images in this way can allow creases and folds in the packaging material to be spotted as well as seal area contamination. Images in the ultraviolet spectrum are able to detect structural changes in thermoplastic long after the seal has been made, so if the packaging material structures have been heat damaged during the sealing process it will be seen by the cameras. The use of high-resolution images in the visible spectrum can allow for detection of small areas of contamination as well as other faults.

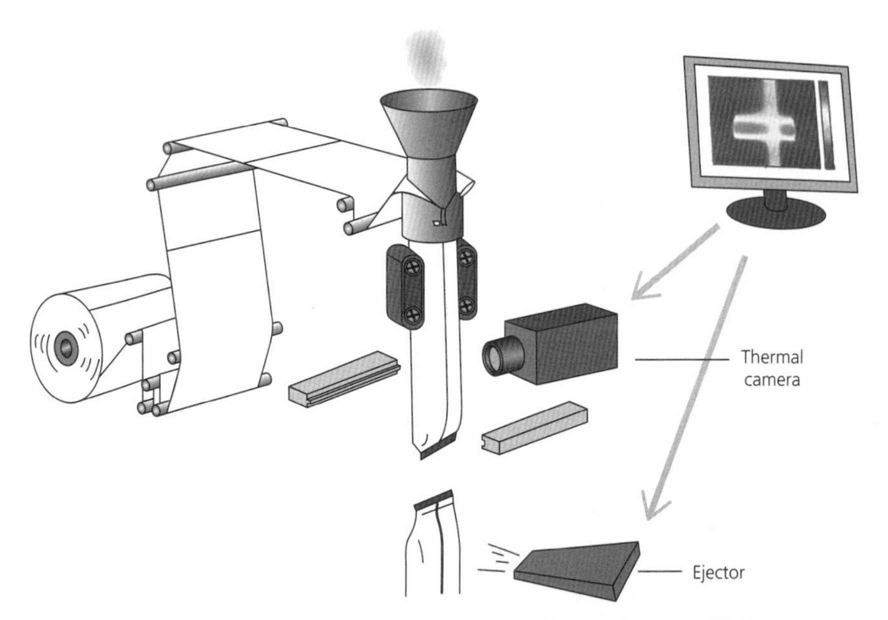

Fig. 14.1 Multispectral cameras are now able to inspect under a wide range of lighting frequencies and consequently detect things that are invisible to the human eye.

The speed of image analysis is always an issue on high-speed packing lines and this problem can be reduced by only analysing the image in the seal area. This greatly reduces the quantity of processing required and therefore speeds up the image analysis time. Computers are getting faster and faster all the time and the resolution of the images and lens systems are also developing at the same pace. Image analysis is being developed into a full replacement for 'visual inspection' by an operator and will allow full automation of the case packing operation without the risk of faulty pack seals getting out into the supply chain.

14.4.5 Laser scatter

This is a subset of image analysis that relies on a laser light source in the form of a laser line being reflected from a seal area. The laser is scattered during the reflection and this technique measures the quantity of laser scattering and the angle of the scattered beam. An anomaly in the seal area will cause the laser line to scatter more than normal as the pack moves along the conveyor and under the laser source. The degree of scatter can be analysed and decisions made about the quality of the seals (Fig. 14.2).

14.4.6 Polarised light

This is another subset of image analysis, where the pack is illuminated using polarised light, either by transmission through the seal area or reflection off the surface (Fig. 14.3a). The camera has a polarising filter and so stress patterns in the

Fig. 14.2 A laser scatter image that shows the patterns seen when a laser light source is reflected off the seal area of a package that contains a seal fault (top) and an acceptable seal (bottom).

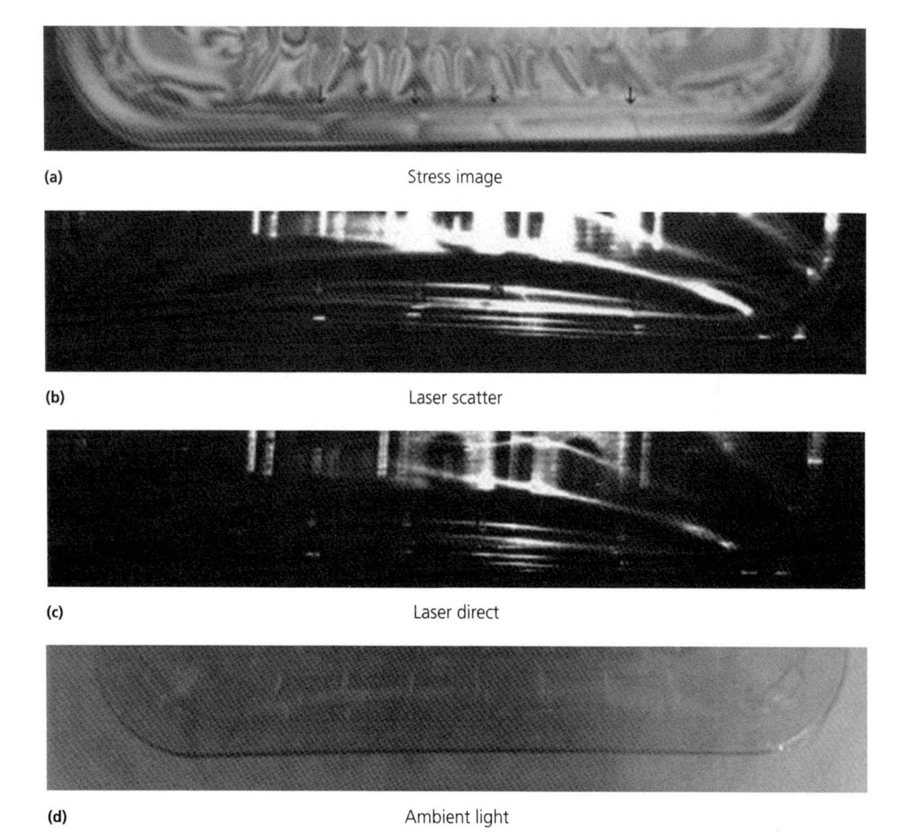

Fig. 14.3 Polarised light transmitted through a seal area on a tray with four known faults (arrowed). The faults are invisible in the ambient light image **(d)** but are highlighted in the polarised light image **(a)** and the two laser images **(b)** and **(c)**.

films can be identified. The polarised image can be analysed and decisions taken about the quality of the seals. Dark fringes that emanate from a point on the seal area are a strong indication of a seal anomaly.

14.4.7 Multiple image analysis

Each of the image analysis systems above have an error where false positives or false negatives can occur. The use of more than one technique in parallel can greatly increase the accuracy of the information produced. So, for example, the laser scatter system may be 80% accurate and the polarised light system 85% accurate for a fault of 10 or 15 microns in size. Used together they may well be able to identify 95 or 96% of the pack faults because the systems are so different and are looking at different aspects of the same seal. This additionality is a major area for development of on-line seal inspection and quality measurement systems.

14.4.8 Active and intelligent inks

There is a lot of work currently being carried out looking at the use of inks that change in some way to add additional functionality to a food package: colour-changing inks that are sensitive to temperature or the condition of the food inside the pack, or inks that have an effect over and above their normal function. Clear inks that absorb a high proportion of ultraviolet light are an example of the work being done. This ink opens up the possibility of meeting the demands of the consumer to be able to see the product at the same time as protecting the product from the harmful effects of ultraviolet light on colour and flavour compounds in the foods.

14.4.9 Bump Mark

This is an award-winning innovation that uses materials that are sensitive to temperature and the length of time for which a food is stored. As the food goes through its shelf life so does the label – at exactly the same rate. So no need for 'use by' dates on the package; the food is safe to eat if the label is smooth to touch. If the label feels bumpy, the product is getting too old. This is great for those with visual impairments as well as the general public. The Bump Mark could also reduce the amount food wasted by those who use the dates on packaging as a 'throw it away date'.

14.4.10 Smart labels

This innovation is really in the printing of electrical circuits in a way that greatly reduces the cost of electrical sensors and other devices. It will soon be possible to use your smart phone to get much more information from packaging than is currently available – even with very small printing. The smart labels will enable the analysis and communication of information by the packaging. How easy will it be to communicate with the person who ate a ready meal from your factory when you have GPS data as well as their email address?

Index

Handbook of Seal Integrity in the Food Industry, First Edition. Michael Dudbridge.